THE
INDUCTION
BOOK

A U R O R A
DOVER MODERN MATH ORIGINALS

Dover Publications is pleased to announce the publication of its first volumes in our new Aurora Series of original books in mathematics. In this series, we plan to make available exciting new and original works in the same kind of well-produced and affordable editions for which Dover has always been known.

Aurora titles currently available are:

Numbers: Histories, Mysteries, Theories by Albrecht Beutelspacher. (978-0-486-80348-7)

Algebra: Polynomials, Galois Theory and Applications by Frédéric Butin. (978-0-486-81015-7)

An Interactive Introduction to Knot Theory by Inga Johnson and Allison K. Henrich. (978-0-486-80463-7)

The Theory and Practice of Conformal Geometry by Steven G. Krantz. (978-0-486-79344-3)

Category Theory in Context by Emily Riehl. (978-0-486-80903-8)

Optimization in Function Spaces by Amol Sasane. (978-0-486-78945-3)

An Introductory Course on Differentiable Manifolds by Siavash Shahshahani. (978-0-486-80706-5)

Calculus: A Rigorous First Course by Daniel J. Velleman. (978-0-486-80936-6)

Elementary Point-Set Topology: A Transition to Advanced Mathematics by André L. Yandl and Adam Bowers. (978-0-486-80349-4)

Additional volumes will be announced periodically.

The Dover Aurora Advisory Board:

John B. Little
College of the Holy Cross
Worcester, Massachusetts

Daniel S. Silver
University of South Alabama
Mobile, Alabama

THE
INDUCTION
BOOK

STEVEN H. WEINTRAUB
Lehigh University

DOVER PUBLICATIONS, INC.
MINEOLA, NEW YORK

Bibliographical Note

The Induction Book is a new work, first published by Dover Publications, Inc., in 2017.

International Standard Book Number
ISBN-13: 978-0-486-81199-4
ISBN-10: 0-486-81199-9

Manufactured in the United States by LSC Communications
81199901   2017
www.doverpublications.com

Contents

CONTENTS

Preface

At its heart, this is a problem book about mathematical induction.

Why a book about induction? The answer is simple but compelling.

Mathematical induction, its equivalents complete induction and well-ordering, and its immediate consequence, the pigeonhole principle, are important proof techniques in mathematics. Indeed, they are not only important, but essential and ubiquitous. Every mathematician is familiar with mathematical induction, and every student of mathematics needs to be. Thus we have written this book to provide the reader with an introduction and a thorough exposure to these proof techniques.

To whom is it addressed? There are several audiences.

1. This book is well suited to be used for a course on mathematical induction. The author has used parts of this book for such a course. There is far more material in this book than can be covered in such a course, so instructors may pick their favorite topics from among the ones presented here.

2. Most theorem-proving courses include a segment on induction. Thus this book can be used as a supplement for such courses, providing additional explanation and additional problems to be solved.

3. Since this book contains a large collection of problems, it can be used in problem-solving courses. The author has often taught problem-solving courses ("coaching" for the Putnam competition), using some of these problems in these courses.

4. Students looking for interesting and challenging problems to cut their teeth on will find a variety of them here.

Let us next describe the plan of the book.

Despite its power, the basic idea of mathematical induction is quite simple. Thus we begin in Chapter 1 with an intuitive explanation of mathematical induction and its equivalents, and then proceed to formalize it. This basic idea appears in many variants, so we give a number of illustrative examples of its use.

The core of this book is Chapter 2, a large collection of problems consisting of results to be proved by induction or by the pigeonhole principle. (Henceforth when we say induction, we mean mathematical induction, complete induction, or well-ordering). These problems are deliberately presented *without* solutions to enable instructors to assign them to their students (and to keep students who read this book on their own honest).

A typical induction problem is a two-step problem. The first step is to find a pattern, and the second step is to prove that it holds. Induction is a proof technique, not a discovery technique, so it applies to the second step, not the first.

Many of our problems just involve the second step: We give a result, and the problem is to find a proof of it. Some involve both steps: the reader must first discover the pattern, and then prove that it holds. Some of our problems are relatively straightforward, while others require varying degrees of cleverness, ingenuity, and hard work. The author of this book thinks of problem-solving as fun. We hope the readers of this book will have fun attacking the problems here.

Our problems range from "old chestnuts" through original problems. Some of these are problems that are interesting but not important in themselves, being chosen to give the reader practice in proofs by induction. Others are important theorems that can be proved by induction. While we have mentioned that induction is ubiquitous in mathematics, we restrict ourselves here to elementary problems. By this we mean that most problems require little background, only material that any college (or advanced high school) student should know. There are some problems that involve calculus or linear algebra, but none more advanced than that. There are many problems that involve elementary number theory, as many of the basic theorems of number theory can be proved by induction. (For some of the number theory problems and results, the reader should be familiar with congruences.)

An inductive argument is often the acorn from which a mighty mathematical oak grows. Thus in some of those instances we have followed this growth, expanding on the mathematics in order to illustrate the consequences of the inductive argument (thereby showing the power of induction).

To say that a problem is elementary is not to say that it is easy. The problems here have varying levels of difficulty. There are some beautiful and important theorems that can be proved by induction, but whose proofs are just too difficult to expect students to be able to find on their own. Thus we include an exposition of some of these theorems and their proofs in Chapter 3. (Among these are some famous theorems by famous mathematicians.)

Finally, we have mentioned that mathematical induction, complete induction, and well-ordering are logically equivalent. The reader can take this on faith, and use whichever of these is most convenient in solving a particular problem. But taking things on faith is not a satisfactory way to proceed in mathematics, in the end, so we include an appendix proving this equivalence.

The reader of this book will learn some interesting and beautiful mathematics along the way, and it is one of our goals to present this. We even think that most professional mathematicians will find some items with which they were previously unfamiliar, so they should enjoy it, too.

We make some remarks about notation and language. Results in this book have three-level numbering, so that, for example, Theorem 3.1.9 is the 9th numbered item in Chapter 3, Section 1. The ends of proofs are marked by the symbol □. In some cases we follow the statements of problems with additional remarks. In order to make clear where the problems themselves end, we mark the ends of all problems by the symbol ◇.

We recall that, for any function $f(k)$, $\sum_{k=1}^{n} f(k)$ is the sum $f(1) + f(2) + \ldots + f(n)$ and $\prod_{k=1}^{n} f(k)$ is the product $f(1)f(2)\cdots f(n)$. We follow the standard convention that the empty sum is 0 and the empty product is 1, so that, for example, $\sum_{k=1}^{0} f(k) = 0$ and $\prod_{k=1}^{0} f(k) = 1$. We also recall that, as an analytic expression, x^y is undefined whenever $x = y = 0$. However, in this book whenever this situation comes up we will always be dealing with symbolic or combinatorial expressions, rather than analytic ones, and so we will understand that $0^0 = 1$ throughout.

Several of the problems in this book are modifications of problems that have appeared in Putnam competitions and the author thanks the Mathematical Association of America for permission to use them.

Finally, this book was begun while the author was physically at Lehigh, and was completed while the author was on sabbatical leave at the Mathematics Institute of the University of Göttingen. He thanks Lehigh for the time off to finish the book, and the Mathematics Institute for its hospitality during his visit.

Steven H. Weintraub
Bethlehem, PA USA
March, 2016

Chapter 1

Introducing Induction

In this chapter we introduce mathematical induction, first informally, and then formally. We then introduce complete induction and well-ordering, and derive the pigeonhole principle. Finally, we give a variety of examples of proofs using these methods.

1.1 The Principle of Mathematical Induction

Suppose we have a row of dominoes, with each domino labeled by a positive integer. Suppose also that the dominoes are arranged so that

(a) If domino n falls, it knocks over domino $n + 1$.

Now suppose that

(b) The first domino falls.

What will be the result? Clearly it will be that

All the dominoes fall.

This is nothing more or less than the principle of mathematical induction.

Let us formalize this. Let $D(n)$ be the statement that domino n falls. Condition (a) is the statement that

If $D(n)$ is true, then $D(n + 1)$ is also true,

while condition (b) is the statement that

D(1) is true,

and our conclusion is the statement that

D(n) is true for every positive integer n.

Once we view things in this way, we see there is nothing special about dominoes. By exactly the same logic, we have:

Axiom 1.1.1 (The Principle of Mathematical Induction). *Let $P(n)$ be any proposition about the positive integer n. If*

(1) $P(1)$ is true; and

(2) If $P(n)$ is true, then $P(n+1)$ is true;

then

$P(n)$ is true for every positive integer n.

In this statement, a "Proposition" is simply any true-false statement. Also, we have reversed the order of conditions (a) and (b) in our discussion of dominoes in stating conditions (1) and (2) in this statement to conform with usual mathematical practice. Condition (1) is often called the *base case*, while condition (2) is often called the *inductive step*, and we shall use this language.

To emphasize, the inductive step is a conditional one, and that is its power: We are allowed to assume that $P(n)$ is true, and can use that assumption to prove that $P(n+1)$ is true.

As a matter of practice, it is usually easy to verify that the base case is true, and the work comes in showing that the inductive step is true. Occasionally, it is easy to show that the inductive step is true, and what requires work is to verify the base case. Very rarely, we have to work at both (and almost never are both easy–that would be getting something for nothing).

Here is almost everybody's first example of a proof by induction.

Theorem 1.1.2. *Let S_n be the sum of the first n positive integers, $S_n = \sum_{i=1}^{n} i$. Then $S_n = n(n+1)/2$.*

Proof. We prove this by induction. Let $P(n)$ be the proposition:

$$S_n = \frac{n(n+1)}{2}.$$

The base case: We observe that $S_1 = 1$, and that, for $n = 1$, $n(n+1)/2 = 1(2)/2 = 1$, and these are equal. Thus $P(1)$ is true.

The inductive step: We observe that, by definition,

$$S_{n+1} = \sum_{i=1}^{n+1} i = \left(\sum_{i=1}^{n} i\right) + (n+1)$$

$$= \frac{n(n+1)}{2} + (n+1) \quad \text{by the inductive hypothesis}$$

$$= \frac{n(n+1)}{2} + \frac{2(n+1)}{2}$$

$$= \frac{(n+2)(n+1)}{2}$$

$$= \frac{(n+1)((n+1)+1)}{2}$$

and thus $P(n+1)$ is true.

Then, by induction, we conclude that $P(n)$ is true for every positive integer n, i.e., that $S_n = n(n+1)/2$ for every positive integer n. □

Note that in the above proof, we have used the phrase *by the inductive hypothesis* at the point where we have used the fact that $P(n)$ is true. This is standard mathematical practice, and we will always follow it here. It is sometimes omitted in more advanced mathematical work, when the reader is expected to be able to figure that out for herself or himself. I strongly recommend that you always include it in your proofs to sharpen your own logic: You need to clearly see at exactly what point in the proof you use the fact that $P(n)$ is true.

You may legitimately ask how we arrived at the above formula for S_n. The answer is that someone first figured it out. Remember that mathematical induction is a method of proof, not of discovery.

Here is a second proof of this theorem:

Proof. For any n, $S_n = 1+2+3+\ldots+(n-2)+(n-1)+n$. But then also $S_n = n+(n-1)+(n-2)+\ldots+3+2+1$. Let us add these expressions:

$$S_n = 1 + 2 + 3 + \ldots + (n-2) + (n-1) + n$$
$$S_n = n + (n-1) + (n-2) + \ldots + 3 + 2 + 1$$

$$2S_n = (n+1) + (n+1) + (n+1) + \ldots + (n+1) + (n+1) + (n+1)$$
$$= n(n+1) \quad \text{as there are } n \text{ columns,}$$

so

$$S_n = \frac{n(n+1)}{2}.$$ □

In this proof, we see a clever trick. Given this trick, we can come up with the formula for S_n, not just prove it.

But we should point out that while this proof *seems* not to use induction, in fact, logically speaking, it *does*. Where does it do so? The induction is hidden in the ellipses (...). What do the ellipses mean? It means there is a pattern, and we can *see* that this pattern continues for every n. But how can we *prove* that it does? That requires induction.

Here is yet another proof, which proceeds by seeing a different pattern and proving, by induction, that it is always true. We are trying to figure out a formula for S_n, so let us make a table of values of S_n for small values of n:

n	S_n
1	1
2	3
3	6
4	10
5	15
6	21

We stare at this table long enough, and then we see: $S_n + S_{n+1}$ is always a perfect square!

n	S_n	$S_n + S_{n+1}$
1	1	4
2	3	9
3	6	16
4	10	25
5	15	36
6	21	49

In fact, we see that $S_n + S_{n+1} = (n+1)^2$. So we shall prove this is true by induction, and then use it to derive our formula for S_n.

Theorem 1.1.3. *Let S_n be the sum of the first n positive integers, $S_n = \sum_{i=1}^{n} i$. Then $S_n + S_{n+1} = (n+1)^2$.*

Proof. We prove this by induction. Let $P(n)$ be the proposition:

$$S_n + S_{n+1} = (n+1)^2.$$

The base case: We observe that $S_1 = 1$ and $S_2 = 3$, so that $S_1 + S_2 = 4$ and that, for $n = 1$, $(n+1)^2 = 2^2 = 4$, and these are equal. Thus $P(1)$ is true.

The inductive step: We observe that

$$S_{n+1} = \sum_{i=1}^{n+1} i = \left(\sum_{i=1}^{n} i\right) + (n+1) = S_n + (n+1)$$

and similarly

$$S_{n+2} = \sum_{i=1}^{n+2} i = \left(\sum_{i=1}^{n+1} i\right) + (n+2) = S_{n+1} + (n+2).$$

Then

$$
\begin{aligned}
S_{n+1} + S_{n+2} &= (S_n + (n+1)) + (S_{n+1} + (n+2)) \\
&= (S_n + S_{n+1}) + ((n+1) + (n+2)) \\
&= (n+1)^2 + ((n+1) + (n+2)) \quad \text{by the induction hypothesis} \\
&= (n^2 + 2n + 1) + (2n + 3) = n^2 + 4n + 4 \\
&= (n+2)^2 = ((n+1) + 1)^2
\end{aligned}
$$

and thus $P(n+1)$ is true.

Then, by induction, we conclude that $P(n)$ is true for every positive integer n, i.e., that $S_n + S_{n+1} = (n+1)^2$ for every positive integer n. $\square$

Corollary 1.1.4. *Let S_n be the sum of the first n positive integers, $S_n = \sum_{i=1}^{n} i$. Then $S_n = n(n+1)/2$.*

Proof. We have just proved that $S_n + S_{n+1} = (n+1)^2$ for every positive integer n, and in the course of that proof, we observed that $S_{n+1} = S_n + (n+1)$ for every positive integer n. We then have

$$
\begin{aligned}
S_n + S_{n+1} &= (n+1)^2 \\
S_n + (S_n + (n+1)) &= (n+1)^2 \\
2S_n + (n+1) &= (n+1)^2 \\
2S_n &= (n+1)^2 - (n+1) = n(n+1) \\
S_n &= n(n+1)/2
\end{aligned}
$$

as claimed. $\square$

As a second sort of example, we turn to a family of incredibly useful algebraic identities. We can easily verify the following:

$$x - y = (x - y)1$$
$$x^2 - y^2 = (x - y)(x + y)$$
$$x^3 - y^3 = (x - y)(x^2 + xy + y^2)$$
$$x^4 - y^4 = (x - y)(x^3 + x^2 y + xy^2 + y^3)$$

We see a clear pattern emerging, which leads us to make the following conjecture. (In stating this conjecture, we understand that $x^0 = 1$ and $y^0 = 1$.)

Conjecture 1.1.5. *For any positive integer n,*

$$x^n - y^n = (x - y) \sum_{i=0}^{n-1} x^{n-1-i} y^i.$$

Informally, we can prove this by multiplying out:

$$x^{n-1} + x^{n-2}y + x^{n-3}y^2 + \ldots + x^2 y^{n-3} + xy^{n-2} + y^{n-1}$$
$$x \qquad - y$$

$$\overline{\qquad\qquad\qquad\qquad\qquad\qquad\qquad\qquad\qquad\qquad\qquad}$$

$$- x^{n-1}y - x^{n-2}y^2 - x^{n-3}y^3 - \ldots - x^2 y^{n-2} - xy^{n-1} - y^n$$
$$x^n + x^{n-1}y + x^{n-2}y^2 + x^{n-3}y^3 + \ldots + x^2 y^{n-2} + xy^{n-1}$$

$$\overline{\qquad\qquad\qquad\qquad\qquad\qquad\qquad\qquad\qquad\qquad\qquad}$$

$$x^n \qquad\qquad\qquad\qquad\qquad\qquad\qquad\qquad\qquad - y^n$$

But again, because of the ellipses, this is not a rigorous proof. We now present one, by induction (of course).

Theorem 1.1.6. *For any positive integer n,*

$$x^n - y^n = (x - y) \sum_{i=0}^{n-1} x^{n-1-i} y^i.$$

Proof. Let $P(n)$ be the statement of the theorem for a particular value of n.

The base case: For $n = 1$, $x^n - y^n = x^1 - y^1 = x - y$ and $(x - y) \sum_{i=0}^{n-1} x^{n-1-i} y^i = (x - y) \sum_{i=0}^{0} x^{n-1-i} y^i = (x - y)x^0 y^0 = (x - y)1 = (x - y)$ and these are equal, so $P(1)$ is true.

The inductive step: We use the (common) trick of adding and subtracting the same quantity:

$$x^{n+1} - y^{n+1}$$
$$= x^{n+1} - x^n y + x^n y - y^{n+1}$$
$$= (x - y)x^n + y(x^n - y^n)$$
$$= (x - y)x^n + y\left((x - y)\sum_{i=0}^{n-1} x^{n-1-i}y^i\right) \quad \text{by the inductive hypothesis}$$
$$= (x - y)x^n + (x - y)\sum_{i=0}^{n-1} x^{n-(i+1)}y^{i+1}$$
$$= (x - y)x^n y^0 + (x - y)\sum_{j=1}^{n} x^{n-j}y^j \quad (\text{setting } j = i + 1)$$
$$= (x - y)\sum_{j=0}^{n} x^{n-j}y^j$$

and thus $P(n + 1)$ is true.

Hence, by induction, we conclude that $P(n)$ is true for every positive integer n. $\qquad \square$

This formula is true as an algebraic expression, so it remains true if we substitute numerical values for x and y. We used $x^0 = 1$ and $y^0 = 1$ so strictly speaking, we must only substitute nonzero values of x and y. But that was just for convenience, and the formula remains true without this restriction. If $y = 0$ it is the identity $x^n = x(x^{n-1})$). If $x = 0$ it is the identity $-y^n = -y(y^{n-1})$). If $x = y = 0$ it is the identity $0 = 0$.

As a corollary of this formula we obtain an expression for the sum of the first n terms of the geometric progression $a, ar, \ldots, ar^{n-1}$.

Corollary 1.1.7. *The sum of the first n terms of the geometric progression* $a, ar, \ldots, ar^{n-1}$ *is*

$$\sum_{i=0}^{n-1} ar^i = na \quad \text{if } r = 1,$$

$$= a\frac{1 - r^n}{1 - r} \quad \text{if } r \neq 1.$$

Proof. If $r = 1$, then this sum is just $\sum_{i=0}^{n-1} a = na$.

Suppose $r \neq 1$. Substituting $x = 1$ and $y = r$ in the above formula, we see that $1 - r^n = (1 - r) \sum_{i=0}^{n-1} r^i$ and so $\sum_{i=0}^{n-1} r^i = (1 - r^n)/(1 - r)$. But then

$$\sum_{i=0}^{n-1} ar^i = a \sum_{i=0}^{n-1} r^i = a\frac{1 - r^n}{1 - r}$$

as claimed.   □

We conclude this section by mentioning a slight invariant of mathematical induction. Most of the time, we want to use mathematical induction to prove propositions about the positive integers, as we have just done. But it is not uncommon to want to use mathematical induction to prove propositions about the nonnegative integers, and occasionally we want to use mathematical induction to prove propositions about integers greater than or equal to a fixed integer n_0. But the logic is exactly the same. Referring back to the idea of dominoes, instead of numbering the first domino 1, we number the first domino n_0. Then when that domino falls, and each domino knocks down the next one, we obtain that every domino n with $n \geq n_0$ falls. (The point is, we need some domino to start with, but it doesn't matter which one that is.)

Thus we have the following variant of mathematical induction, which we call by the same name.

Axiom 1.1.8 (The Principle of Mathematical Induction). *Let n_0 be any fixed integer. Let $P(n)$ be any proposition about the integer $n \geq n_0$. If*

(1) $P(n_0)$ is true; and

(2) For $n \geq n_0$, if $P(n)$ is true, then $P(n+1)$ is true;

then

   $P(n)$ is true for every integer $n \geq n_0$.

(Then, in particular, the case of positive integers corresponds to $n_0 = 1$, and the case of nonnegative integers corresponds to $n_0 = 0$.)

However, we wish to emphasize that there needs to be a starting value, and for any value there needs to be a next value. Thus, mathematical induction does not directly apply to prove propositions about all integers, as there is no smallest value to start with (although given any integer, there is a next one), nor does it directly apply to prove propositions about all nonnegative rational numbers, since (even though there is a starting value, namely 0), given a rational number, there is no next one. (We have written "directly apply" as sometimes we can be clever enough to modify the argument in order to be able to apply induction.)

1.2 The Principle of Complete Induction

Complete induction is a variant of mathematical induction. To explain it, we return to dominoes.

Again suppose we have a row of dominoes, with each domino labeled by a positive integer. Suppose, for example, that the dominoes are arranged as follows: Domino 2 is knocked down by domino 1. Domino 3 is knocked down by domino 2. Domino 4 is also knocked down by domino 2. Domino 5 is a heavy domino, so is not knocked down by any individual domino, but is knocked down by dominoes 3 and 4 together. And in general, we suppose that domino $n + 1$ is knocked down by some perhaps mysterious combination of dominoes 1 through n, for each n. Now domino 1 falls. What will be the result? Again it will be that all the dominoes fall. This is the principle of complete induction.

In thinking how to state it, we see that it might be difficult to specify exactly which combination of dominoes 1 through n knocks down domino $n + 1$, so we make life easy on ourselves: We just suppose that if all dominoes 1 through n fall, they take down domino $n + 1$ with them. With this in mind, we formulate the principle of complete induction. However, bearing in mind our last formulation of the principle of mathematical induction, where we wished to have the freedom to start with any integer n_0, not just with $n_0 = 1$, we formulate this principle accordingly.

Axiom 1.2.1 (The Principle of Complete Induction). *Let n_0 be any fixed integer. Let $P(n)$ be any proposition about the integer $n \geq n_0$. If*

(1) $P(n_0)$ is true; and

(2) If $P(k)$ is true for all values of k with $n_0 \leq k \leq n$, then $P(n + 1)$ is true;

then

      $P(n)$ is true for every integer $n \geq n_0$.

In fact, complete induction is logically equivalent to mathematical induction. We defer the proof of this to the appendix, and focus here on how to use it.

Consider the sequence defined as follows:

$$x_0 = a, \, x_1 = b, \, \text{ and } x_{n+2} = \frac{x_n x_{n+1}}{2x_n - x_{n+1}} \text{ for } n \geq 0$$

where a and b are positive real numbers with $a > b$. We wish to find and prove an expression for x_n in general.

We begin by experimenting to see if we can find a pattern:

$$x_0 = a$$
$$x_1 = b$$
$$x_2 = \frac{ab}{2a - b}$$

$$x_3 = \frac{b\frac{ab}{2a-b}}{2b - \frac{ab}{2a-b}} = \frac{ab}{3a - 2b}$$

$$x_4 = \frac{\frac{ab}{2a-b}\frac{ab}{3a-2b}}{2\frac{ab}{2a-b} - \frac{ab}{3a-2b}} = \frac{ab}{4a - 3b}$$

(Here we have skipped some complicated but elementary algebraic steps.) At this point a pattern suggests itself:

$$x_n = \frac{ab}{na - (n-1)b}.$$

Thus we are led to conjecture the following result:

Proposition 1.2.2. *Consider the sequence:*

$$x_0 = a, x_1 = b, \text{ and } x_{n+2} = \frac{x_n x_{n+1}}{2x_n - x_{n+1}} \text{ for } n \geq 0$$

where a and b are positive real numbers with $a > b$. Then

$$x_n = \frac{ab}{na - (n-1)b} \text{ for every } n \geq 0.$$

Proof. By induction. Let $P(n)$ be the claim $x_n = ab/(na - (n-1)b)$.

The base cases: For $n = 0$, $ab/(na - (n-1)b) = ab/(-(-b)) = a$, so $P(0)$ is true. Also, for $n = 1$, $ab/(na - (n-1)b) = ab/a = b$, so $P(1)$ is true.

The inductive step: By definition,

$$
\begin{aligned}
x_{n+1} &= \frac{x_{n-1}x_n}{2x_{n-1} - x_n} \\[2mm]
&= \frac{\frac{ab}{(n-1)a-(n-2)b}\frac{ab}{na-(n-1)b}}{2\frac{ab}{(n-1)a-(n-2)b} - \frac{ab}{na-(n-1)b}} \quad \text{by the inductive hypothesis} \\[2mm]
&= \frac{(ab)(ab)}{2ab\,((na-(n-1)b)) - ab\,((n-1)a - nb))} \\[2mm]
&= \frac{a^2b^2}{a^2b\,(2n-(n-1)) - ab^2\,(2(n-1)-(n-2))} \\[2mm]
&= \frac{ab}{(n+1)a - nb}
\end{aligned}
$$

so $P(n+1)$ is true. Hence, by complete induction, $P(n)$ is true for every $n \geq 0$. ☐

There are two important things to notice about this proof. First, we used complete induction rather than induction, since we needed to know that *both* $P(n-1)$ and $P(n)$ were true (not just $P(n)$) in order to conclude than $P(n+1)$ is true. Second, we needed to verify *two* base cases here (not just one). To see why this is true, let us examine the argument a little more carefully. Suppose we had only verified $P(0)$ as our base case. In order to apply the inductive step in the first case, to conclude that $P(2)$ is true, we would have to have known that both $P(0)$ is true and $P(1)$ is true. We would have known the first of these, but not the second, and so we would have been stuck. But once we knew both $P(0)$ and $P(1)$, we were able to conclude $P(2)$, and then we knew both $P(1)$ and $P(2)$, so we were able to conclude $P(3)$, etc., and our proof works.

1.3 The Well-Ordering Principle

The well-ordering principle is a third variant of mathematical induction. Its statement is more technical and less obvious. But it turns out that this more technical formulation is often more useful, so we give it here, and use it. Again we state it in a way that allows us to begin with any integer n_0.

Axiom 1.3.1 (The Well-Ordering Principle). *Let n_0 be any fixed integer. The set $T = \{n \geq n_0\}$ of integers greater than or equal to n_0 is well-ordered, i.e., any nonempty subset S of T has a least element.*

(For $n_0 = 1$ (respectively, $n_0 = 0$), this is the statement that the set of positive integers (respectively, the set of nonnegative integers) is well-ordered.)

Again we defer the proof that the well-ordering principle is logically equivalent to mathematical induction (or complete induction) to the appendix, and concentrate here on using it.

This is certainly not as intuitively clear as the principle of mathematical induction, but let us see how to apply it to dominoes. Suppose we have our usual arrangement of dominoes numbered $1, 2, 3, \ldots$ where, if domino n falls, it knocks down domino $n + 1$, and domino 1 falls. We will use well-ordering to show that all the dominoes fall. (This is a typical example of a proof by contradiction.) Suppose not. Let S be the set of dominoes that do not fall. We are supposing that not all the dominoes fall, so S is nonempty. Then, by well-ordering, S has a smallest element s_0. What could s_0 be? It can't be 1 because we know that domino 1 falls. Thus $s_0 > 1$. Now s_0 is the smallest element of S, so $s_0 - 1$ is not in S, i.e., domino $s_0 - 1$ falls. But then domino s_0 also falls. This means that s_0 is *not* in S, which is a contradiction. Thus our supposition that not all the dominoes fall leads to a contradiction, so that supposition must be false, and we conclude that indeed all the dominoes do fall.

Let us draw an immediate, but important, consequence of well-ordering.

Theorem 1.3.2. *Any strictly decreasing sequence $x_1 > x_2 > x_3 > \ldots$ of positive integers must be finite.*

Proof. Let $S = \{x_1, x_2, x_3, \ldots\}$ be the set of elements of the sequence. Then, by the well-ordering principle, S has a smallest element x_k for some k. But then the sequence must end with x_k, as if it continued, we would have $x_{k+1} < x_k$, contradicting the minimality of x_k.   □

Of course, it is not true that every nonempty subset S of the set of positive integers has a largest element as $S = \{1, 2, 3, \ldots\}$, the set of all positive integers, does not. It is also not true that any strictly increasing sequence of positive integers must be finite, as the sequence $1 < 2 < 3 < \ldots$ of all positive integers is not. But if the set (or sequence) is bounded from above, the story changes.

Theorem 1.3.3. *(a) Let S be any nonempty subset of the set of positive integers that is bounded from above, i.e., for which there exists a positive integer N such that $s < N$ for every element s of S. Then S has a largest element.*

(b) Any strictly increasing sequence $x_1 < x_2 < x_3 < \ldots$ of positive integers that is bounded from above has a largest element.

Proof. (a) Let S' be the set $S' = \{N - s \mid s \in S\}$. Then, by well-ordering, S' has a smallest element s'. But then $s' = N - s$ for some $s \in S$, in which case $s = N - s'$ is the largest element of S.

(b) Suppose that $x_1 < x_2 < x_3 < \ldots$ is a strictly increasing sequence of positive integers with $x_i < N$ for some N, for every i. Then, setting $y_i = N - x_i$, we see that $y_1 > y_2 > y_3 > \ldots$ is a strictly decreasing sequence of positive integers, so it must be finite. Let this sequence end at y_k. Then the original sequence ends at x_k.   □

The well-ordering principle is often used to prove that a proposition $P(n)$ about positive integers is always true, by contradiction: Suppose not, then by well-ordering there must be a minimal case in which it is false. But the proposition must be true in that case; contradiction. (This is the sort of argument we gave for dominoes, above.)

However, we will give an example of using it directly to prove an important result.

Theorem 1.3.4 (The division algorithm). *Let a and b be integers, with $b \neq 0$. Then there are unique integers q and r with*

$$a = bq + r \quad with \ 0 \leq r < |b|.$$

Proof. First we prove that there exist integers q and r as claimed.

Let $\varepsilon = 1$ if $b > 0$ and $\varepsilon = -1$ if $b < 0$, so that in any case $|b| = \varepsilon b$.

Let $S = \{$nonnegative integers of the form $a - |b|k$ with k an integer$\}$. We claim S is nonempty. If $a \geq 0$ this is immediate, as we may simply choose $k = 0$. If $a < 0$, choose $k = a$. Then, in this case, $a - a|b| = a(1 - |b|) \geq 0$ (as $a < 0$ and $1 - |b| \leq 0$).

By well-ordering S has a minimal element $r = a - |b|q_0 = a - bq$ where $q = \varepsilon q_0$. We claim $0 \leq r < |b|$. Clearly $0 \leq r$ as r is in S. If $r \geq |b|$, then $r' = a - |b|(q_0 + 1) = r - |b|$ is in S, which is impossible, by the minimality

of r. Thus
$$a = bq + r \quad \text{with } 0 \leq r < |b|.$$

Next we prove that q and r are unique. (This part of the proof does not involve induction.) We do so in the usual way that we show something is unique: We suppose that there are two of them, and show they must be the same.

Thus suppose $a = bq_1 + r_1$ with $0 \leq r_1 < |b|$ and $a = bq_2 + r_2$ with $0 \leq r_2 < |b|$. Subtracting, we see $0 = b(q_1 - q_2) + (r_1 - r_2)r$. Thus $b(q_2 - q_1) = r_1 - r_2$. Call this common value d. Then, on the one hand, the largest value d can take occurs when r_1 is as large as possible, namely $r_1 = |b| - 1$, and when r_2 is as small as possible, namely $r_2 = 0$, so $d \leq (|b| - 1) - 0 = |b| - 1$. Similarly, the smallest value d can take occurs when r_1 is as small as possible, namely $r_1 = 0$, and when r_2 is as large as possible, namely $r_2 = |b| - 1$, so $d \geq 0 - (|b| - 1) = -(|b| - 1)$. Thus $-(|b| - 1) \leq d \leq |b| - 1$. On the other hand, $d = b(q_2 - q_1)$ is divisible by b. But the only number in this range that is divisible by b is $d = 0$. Thus $d = 0$, in which case $r_2 = r_1$, and then also $b(q_2 - q_1) = 0$, so $q_2 = q_1$, and we have uniqueness. □

1.4 The Pigeonhole Principle

In this section we introduce the pigeonhole principle, an intuitively obvious but nevertheless very useful principle. We give several variants of it, and give a typical illustration of its use. Of course, the fact that it is intuitively obvious doesn't mean it doesn't need proof, so we will prove it (by induction, of course).

The name of this principle comes from a venerable technology, but one that is still in use. Suppose a postal employee is sorting n letters into m pigeonholes, and $n > m$ (that is, there are more letters than pigeonholes). What can we conclude? While we cannot say anything about the exact distribution of the letters, we can certainly conclude than at least one pigeonhole must contain more than one letter.

Just as there was nothing special about dominoes, there is nothing special about letters, and so we have the following general principle:

Theorem 1.4.1 (The pigeonhole principle). *If n objects are arranged in m categories, and $n > m$, then some category must contain more than one object.*

In German, the pigeonhole principle is known as Dirichlet's Schubfachprinzip. Schubfachprinzip means desk-drawer principle, and that is easy to understand; we can imagine putting items into desk drawers instead of letters into pigeonholes. But we might wonder why this intuitively obvious principle is named after the famous nineteenth century mathematician Dirichlet. The reason is that he used it to prove an important theorem. His theorem is beyond the scope of this book, but we will give an example of an elementary, but typical, application of the pigeonhole principle.

Magic squares are a staple of recreational mathematics. An n-by-n magic square is an n-by-n square of numbers all of whose rows, columns, and the two diagonals add up to the same number. For example, we have the 3-by-3 magic square:

$$\begin{bmatrix} 8 & 1 & 6 \\ 3 & 5 & 7 \\ 4 & 9 & 2 \end{bmatrix}$$

We also have the 4-by-4 magic square:

$$\begin{bmatrix} 16 & 3 & 2 & 13 \\ 5 & 10 & 11 & 8 \\ 9 & 6 & 7 & 12 \\ 4 & 15 & 14 & 1 \end{bmatrix}$$

(This magic square appears in the famous Albrecht Dürer engraving Melencolia I.)

We are not concerned with magic squares but rather their opposites, antimagic squares. We define an n-by-n *antimagic square* to be an n-by-n square of numbers such that the rows, columns, and two diagonals all add up to distinct numbers. For example, we have the 3-by-3 antimagic square:

$$\begin{bmatrix} 9 & 4 & 5 \\ 10 & 3 & -2 \\ 6 & 9 & 7 \end{bmatrix}$$

We also have the 4-by-4 antimagic square:

$$\begin{bmatrix} 7 & 3 & 6 & 9 \\ 4 & -9 & 5 & 3 \\ 8 & 1 & 3 & 7 \\ 3 & 9 & 0 & 5 \end{bmatrix}$$

Proposition 1.4.2. *For any n, there does not exist an n-by-n antimagic square all of whose entries are -1, 0, or 1.*

Proof. The largest sum a row, column, or diagonal can have is n, if all of its entries are 1. The smallest sum a row, column, or diagonal can have is $-n$, if all of its entries are -1. Thus in any event, the sum of the entries in a row, column, or diagonal must be an integer between $-n$ and n, and there are $2n + 1$ such integers. On the other hand, we have n rows, n columns, and 2 diagonals, for a total of $2n + 2$ sums. Thus, for some m with $-n \leq m \leq n$, there must be two sums that have the same value m, and so we cannot have such an antimagic square. $\square$

This is a typical application of the pigeonhole principle in that we know very little about the individual sums, but can simply count them to prove that they cannot be distinct.

We will now prove a strengthened version of the pigeonhole principle.

Theorem 1.4.3 (The pigeonhole principle, strong form). *(a) If n objects are arranged in m categories, and $n > m$, then some category must contain more than one object.*

(a') If an infinite number of objects are arranged in a finite number of categories, then some category must contain an infinite number of objects.

(b) If n objects are arranged in m categories, and $n < m$, then some category must be empty.

(c) If n objects are arranged in n categories, then the following are equivalent:

(i) Every category contains exactly one object.

(ii) Every category contains at least one object.

(iii) Every category contains at most one object.

Proof. (a) By induction on m, beginning with $m = 1$.

The base case: If $m = 1$, there is only 1 category, so all of the $n > 1$ objects must be in that category, i.e., that category contains more than one object.

The inductive step: Suppose we have $m + 1$ categories. Consider the last category. Let this category contain k objects. If $k > 1$ we are done. If not, the preceding m categories must contain $n - k$ objects. Since $n > m + 1$ and $k \leq 1$, $n - k > m$, so by the inductive hypothesis at least one of these categories must contain more than one object.

Thus by induction (a) is true for every positive integer m.

(a') The proof, by induction on m, is almost identical to that of (a).

The base case: If $m = 1$, there is only 1 category, so all of the objects must be in that category, i.e., that category contains infinitely many objects.

The inductive step: Suppose we have $m + 1$ categories. Consider the last category. If this category contains infinitely many objects we are done. Otherwise it contains only finitely many objects. Then the preceding m categories must contain the remaining objects, and there are infinitely many of them, so by the inductive hypothesis at least one of these categories must contain infinitely many objects.

Thus by induction (a') is true for every positive integer m.

(b) By complete induction on n, beginning with $n = 0$.

The base case: If $n = 0$, there are no objects, so every category must be empty.

The inductive step: Suppose we have $n + 1$ objects to arrange in m categories, with $n + 1 < m$. Consider the last category. If it is empty, we are done. Otherwise, it contains $k \geq 1$ objects. In that case the preceding $m - 1$ categories contain $n - k < m - 1$ objects, so by the inductive hypothesis at least one of those categories must be empty.

Thus by complete induction (b) is true for every nonnegative integer n.

(c) We use (a) and (b) to prove (c).

Suppose (i) is true. Then (ii) and (iii) are both true.

Suppose (i) is false. Then at least one of (ii) and (iii) is false. If (ii) is false, i.e., if some category is empty, then the remaining $n-1$ categories contain n objects, so by (a) some category contains more than one object, and (iii) is false. If (iii) is false, i.e, if some category contains k objects with $k > 1$, then the remaining $n-1$ categories contain $n - k < n - 1$ objects, so by (b) some category is empty, and (ii) is false.

Thus we see that either (i), (ii), and (iii) are all true, or (i), (ii), and (iii) are all false, i.e., these statements are all logically equivalent.   □

1.5 Further Examples of Proofs by Induction

In this section we give another few proofs by induction, each being chosen to illustrate a point.

The first point we wish to illustrate is that sometimes in proofs by induction, part of the problem is to decide what to induct on. Actually, we already saw this in the preceding section, when in our proof of the strengthened pigeonhole principle, we used induction on m to prove part (a) and induction on n to prove part (b). But let us see another example of this here.

Let b be a fixed integer, $b > 1$. We would like to show that every integer n has a unique base b expansion. The most natural thing to induct on is n, and indeed that can be made to work, but it is complicated. Taking $b = 10$, for example, we know that if n has decimal expansion 12463, then $n + 1$ has decimal expansion 12464, while if n has decimal expansion 12499, $n + 1$ has decimal expansion 12500. In the first case, it is easy to get the expansion of $n+1$ from that of n, but in the second case it is not so easy because of the presence of "carries." Thus we look for a proof that avoids this difficulty.

Theorem 1.5.1. *Let b be any integer greater than 1. Then every positive integer n has a unique base b expansion, i.e., every integer n can be written as*

$$n = \sum_{i=0}^{j} c_i b^i$$

for unique integers $c_0, \ldots, c_j$ with $0 \le c_i \le b-1$ for $i = 0, \ldots, j$ and $c_j \ne 0$.

Proof. First we will prove existence and then uniqueness.

We let $E(k)$ be the proposition that there exists such an expansion for every positive integer $n < b^k$, and prove this by induction on k.

The base case: For $k = 1$, if n is a positive integer $n < b$ then n is an integer with $1 \le n \le b-1$ and so n has the base b expansion given by $j = 0$ and $c_0 = n$.

The inductive step: Let n be an integer with $n < b^{k+1}$. If $n < b^k$ then we are done, by the inductive hypothesis. Otherwise, by the division algorithm,

$n = b^k q + r$ with $0 \leq r < b^k$. Since $b^k \leq n < b^{k+1}$, we must have $1 \leq q \leq b - 1$. If $r = 0$ then $n = b^k q$ has the base b expansion given by $j = k$, $c_k = q$, and $c_i = 0$ for $0 \leq i < j$. If $r \neq 0$ then by the inductive hypothesis r has a base b expansion $r = \sum_{i=0}^{j} d_i b^i$, and since $r < b^k$ we have that $j < k$. Then n has the base b expansion given by $j = k$, $c_k = q$, $c_i = 0$ for $j < i < k$, and $c_i = d_i$ for $0 \leq i \leq j$.

Then by induction $E(k)$ is true for every k, so every positive integer n has a base b expansion. (Strictly speaking, in order to conclude this we must show that for any positive integer n, there is a k such that $n < b^k$. But for any positive integer n, $n < b^n$. (Proof by induction: $1 < b = b^1$ and if $n < b^n$, then $n + 1 < b^n + 1 < b^n + b^n(b - 1) = b^{n+1}$.))

We let $U(k)$ be the proposition that the expansion is unique for every positive integer $n < b^k$, and prove this by induction on k.

We suppose we can write $n = \sum_{i=0}^{j'} c_i' b^i$ and $n = \sum_{i=0}^{j''} c_i'' b^i$ and we show that $j'' = j'$ and $c_i'' = c_i'$ for each i.

The base case: Let $n < b^1 = b$. Then we must have $j' = 0$, as otherwise the value of the first sum would be greater than or equal to b, and similarly we must have $j'' = 0$, as otherwise the value of the first sum would be greater than or equal to b. Thus $n = c_0' = c_0''$ and the two expressions agree.

The inductive step: We first observe that for any $j \geq 1$, if $c_i \leq b - 1$ for $0 \leq i \leq j$, then $\sum_{i=0}^{j-1} c_i b^i \leq \sum_{i=0}^{j-1} (b-1) b^i = b^j - 1 < b^j$ by the formula for the sum of a geometric progression.

Let $n < b^{k+1}$. If $n < b^k$, then the expression for n is unique, by the inductive hypothesis. Thus suppose $b^k \leq n < b^{k+1}$. We must have $j' = k$, as if $j' > k$, then $n \geq b^{k+1}$, and if $j' < k$, then $n < b^k$ by the above observation. Similarly $j'' = k$. But then

$$n = b^k c_k' + r' \quad \text{where } r' = \sum_{i=0}^{k-1} c_i' b^i,$$

$$n = b^k c_k'' + r'' \quad \text{where } r'' = \sum_{i=0}^{k-1} c_i'' b^i.$$

Again we observe that $0 \leq r' < b^k$ and $0 \leq r'' < b^k$. But by the uniqueness in the division algorithm we must have $c_k' = c_k''$ and $r' = r''$. But then by the inductive hypothesis we must have that $c_i' = c_i''$ for $0 \leq i \leq k - 1$.

Then by induction $U(k)$ is true for every k, so every positive integer n has a unique base b expansion.   □

The second point we wish to illustrate is that sometimes a proof by induction can actually involve a double induction.

Theorem 1.5.2. *Let n be any positive integer. Then any product of n consecutive positive integers is divisible by $n!$.*

Proof. By induction on n. Let $P(n)$ be the claim in the theorem for n.

The base case: If $n = 1$, it is certainly true that the product of any 1 consecutive positive integers, i.e., any positive integer, is divisible by 1, so $P(1)$ is true.

The inductive step: We let $Q(n + 1, k)$ be the claim that the product of $n + 1$ positive integers beginning with k, i.e., the product $k(k + 1) \cdots (k + n)$, is divisible by $(n + 1)!$. We prove this is true for every positive integer k, by induction on k.

The base case: If $k = 1$, the product $1(2) \cdots (1 + n) = (n + 1)!$ is certainly divisible by $(n + 1)!$, so $Q(n + 1, 1)$ is true.

The inductive step: Consider the product $(k + 1)(k + 2) \cdots (k + 1 + n)$. Then

$$\begin{aligned}
(k + 1)(k + 2) \cdots (k + 1 + n) &= (k + 1)(k + 2) \cdots (k + 1 + n) \\
&\quad - k(k + 1) \cdots (k + n) + k(k + 1) \cdots (k + n) \\
&= ((k + 1)(k + 2) \cdots (k + n)) \, ((k + 1 + n) - k) \\
&\quad + k(k + 1) \cdots (k + n) \\
&= ((k + 1)(k + 2) \cdots (k + n)) \, (n + 1) \\
&\quad + k(k + 1) \cdots (k + n).
\end{aligned}$$

Now $((k + 1)(k + 2) \cdots (k + n))$ is a product of n consecutive integers, so by the inductive hypothesis on n is divisible by $n!$, $((k + 1)(k + 2) \cdots (k + n)) = n!a$ for some integer a. Also, $k(k + 1) \cdots (k + n)$ is the product of $n + 1$ consecutive integers beginning with k, so by the inductive hypothesis on k this product is divisible by $(n + 1)!$, $k(k + 1) \cdots (k + n) = (n + 1)!b$ for some integer b. Substituting, we see

$$(k + 1)(k + 2) \cdots (k + 1 + n) = n!a(n + 1) + (n + 1)!b = (n + 1)!(a + b)$$

is divisible by $(n + 1)!$, and so $Q(n + 1, k + 1)$ is true.

Hence by induction on k, $Q(n + 1, k)$ is true for every positive integer k. But that is the same thing as saying that $P(n + 1)$ is true.

Then, by induction on n, we conclude that $P(n)$ is true for every positive integer n. $\qquad\square$

We should point out that there is a good reason why this theorem is true, and we will see it below. So in this case, the proof we have given is not the most perspicacious one. However, we gave it not because it is the best proof, but rather because we wanted to illustrate the method of double induction.

The third point we want to illustrate is that sometimes in mathematics it is easier to prove a stronger result. For example, suppose we wish to prove

$$\sum_{k=1}^{n} 1/k^2 < 2 \quad \text{for every } n \geq 1.$$

We go to work, by induction. Let $P(n)$ be the claimed result for n.

The base case: For $n = 1$ the claim is $1 < 2$, which is true.
The inductive step: Assume that $P(n)$ is true and consider $P(n+1)$. Then

$$\sum_{k=1}^{n+1} 1/k^2 = \left(\sum_{k=1}^{n} 1/k^2 \right) + 1/(n+1)^2 < 2 + 1/(n+1)^2$$

by the induction hypothesis. But now we are stuck, since we need the right-hand side to be 2, and it isn't.

Thus instead we prove the stronger result

$$\sum_{k=1}^{n} 1/k^2 \leq 2 - 1/n \quad \text{for every } n \geq 1.$$

We go to work, by induction. Let $P(n)$ be the claimed result for n.
The base case: For $n = 1$ the claim is $1 \leq 2 - 1$, which is true.
The inductive step: Assume that $P(n)$ is true and consider $P(n+1)$. Then

$$\sum_{k=1}^{n+1} 1/k^2 = \left(\sum_{k=1}^{n} 1/k^2 \right) + 1/(n+1)^2 < 2 - 1/n + 1/(n+1)^2$$

$$= 2 - (1/n - 1/(n+1)^2)$$

by the inductive hypothesis.

Now we do a little algebra:

$$\frac{1}{n} - \frac{1}{(n+1)^2} = \frac{(n+1)^2 - n}{n(n+1)^2} = \frac{n^2 + n + 1}{(n^2 + n)(n+1)} > \frac{n^2 + n}{(n^2 + n)(n+1)} = \frac{1}{n+1}$$

and so

$$\sum_{k=1}^{n+1} 1/k^2 < 2 - 1/(n+1)$$

so $P(n+1)$ is true. Hence, by induction, $P(n)$ is true for every $n \geq 1$.

A consequence of this result is that $\sum_{k=1}^{\infty} 1/k^2 \leq 2$. It is natural to ask what the value of this sum is. The answer is a famous theorem of Euler.

Theorem 1.5.3 (Euler).

$$\sum_{k=1}^{\infty} \frac{1}{k^2} = \frac{\pi^2}{6}$$

We give this theorem as a matter of mathematical interest, but it is far beyond our means to prove it here.

Chapter 2

Problems

This chapter is a collection of problems to be solved by induction (by which we mean mathematical induction, complete induction or well-ordering) or by the pigeonhole principle. We also include commentary on some of the problems.

These problems include the proofs of important theorems, and in some cases we derive some of their consequences as well.

2.1 Assorted Problems

Problem 2.1.1. Find and prove a formula for value of the sum of the first n odd positive integers

$$\sum_{k=1}^{n} 2k - 1$$

for every positive integer n. ◇

Problem 2.1.2. Show that the value of the sum of the squares of the first n positive integers is given by

$$\sum_{k=1}^{n} k^2 = \frac{n(n+1)(2n+1)}{6}$$

for every positive integer n. ◇

Problem 2.1.3. Show that the value of the sum of the cubes of the first n positive integers is given by

$$\sum_{k=1}^{n} k^3 = \left(\frac{n(n+1)}{2}\right)^2$$

for every positive integer n. ◇

Problem 2.1.4. Show that

$$\sum_{k=1}^{n} \frac{1}{k(k+1)} = \frac{n}{n+1}$$

for every positive integer n. ◇

Problem 2.1.5. Show that

$$\sum_{k=1}^{n} \frac{1}{\sqrt{k}} \leq 2\sqrt{n} - 1$$

for every positive integer n. ◇

Problem 2.1.6. Show that the value of the product of the first n odd positive integers is given by

$$\prod_{k=1}^{n} 2k - 1 = \frac{(2n)!}{2^n n!}$$

for every positive integer n. ◇

Problem 2.1.7. Show that the following algebraic identity holds for every positive integer n:

$$\sum_{k=1}^{n} \left((x_k - 1) \prod_{j=1}^{k-1} x_j \right) = \left(\prod_{k=1}^{n} x_k \right) - 1.$$

For example, the first few cases of this identity are:

$$n = 1: \quad (x_1 - 1) = x_1 - 1$$
$$n = 2: \quad (x_1 - 1) + (x_2 - 1)x_1 = x_1 x_2 - 1$$
$$n = 3: \quad (x_1 - 1) + (x_2 - 1)x_1 + (x_3 - 1)x_1 x_2 = x_1 x_2 x_3 - 1 \qquad ◇$$

Problem 2.1.8. The Fermat number F_n is defined to be $F_n = 2^{2^n} + 1$, for $n \geq 0$. They are so named because Fermat believed that F_n is prime, for every n. Indeed, $F_0 = 3$, $F_1 = 5$, $F_2 = 17$, $F_3 = 257$, and $F_4 = 65537$ are all prime. But Euler discovered that $F_5 = 4294967297$ is divisible by 641, and there is no known value of $n \geq 5$ for which F_n is prime.

(a) Show that

$$\prod_{k=1}^{n-1} F_k = F_n - 2 \quad \text{for every } n \geq 1.$$

More generally, for a fixed integer $m \geq 2$, define $F_n(m)$ by $F_n(m) = m^{2^n} + 1$.

(b) Show that

$$\prod_{k=1}^{n-1} F_k(m) = (F_n(m) - 2)/(m - 1) \quad \text{for every } n \geq 1. \qquad \diamond$$

Problem 2.1.9. A decomposition of a positive integer n is a way of writing n as a sum of positive integers in order. Let $d(n)$ denote the number of decompositions of n. Find, with proof, a formula for $d(n)$. $\qquad \diamond$

By way of contrast, a partition of a positive integer n is a way of writing n as a sum of positive integers, where the order does not matter. Let $p(n)$ denote the number of partitions of n. Then, for example, we have $d(4) = 8$ as 4 has the decompositions 4, $3 + 1$, $1 + 3$, $2 + 2$, $2 + 1 + 1$, $1 + 2 + 1$, $1 + 1 + 2$, $1 + 1 + 1$, while $p(4) = 5$ as 4 has the partitions 4, $3 + 1$, $2 + 2$, $2 + 1 + 1$, $1 + 1 + 1 + 1$. The determination of the value of $p(n)$ for an arbitrary positive integer n is a deep and famous result of Hardy and Ramanujan.

Problem 2.1.10. Show that

$$\sum_{k=1}^{2^n} \frac{1}{k} \geq 1 + n/2$$

for every integer $n \geq 0$. $\qquad \diamond$

Problem 2.1.11. Let a be any fixed real number with $0 < a < 1$. Show that

$$\sum_{k=1}^{2^n} \frac{1}{k^a} \geq 1 + \frac{1}{2} \cdot \frac{2^{(n+1)(1-a)} - 2^{1-a}}{2^{1-a} - 1}$$

for every integer $n \geq 0$. $\qquad \diamond$

Corollary 2.1.12. $\sum_{k=1}^{\infty} \frac{1}{k^a}$ *diverges for $a \leq 1$.*

(Of course, once we know from the preceding problem that this sum diverges for $a = 1$, we know it diverges for $a < 1$ by comparison, but this shows the rate of divergence.)

Problem 2.1.13. Let a be any fixed real number with $a > 1$. Show that

$$\sum_{k=1}^{2^n} \frac{1}{k^a} \leq \frac{1 - 2^{n(1-a)}}{1 - 2^{1-a}}$$

for every integer $n \geq 1$. $\qquad \diamond$

Corollary 2.1.14. $\sum_{k=1}^{\infty} \frac{1}{k^a}$ *converges for a > 1.*

Problem 2.1.15. Let $a_1, a_2, \ldots$ and $b_1, b_2, \ldots$ be arbitrary sequences of complex numbers. Let $A_k = \sum_{i=1}^{k} a_i$. Prove Abel's formula:

$$\sum_{i=1}^{n} a_i b_i = A_n b_n - \sum_{i=1}^{n-1} A_i(b_{i+1} - b_i) \text{ for every } n \geq 1. \qquad \diamond$$

Problem 2.1.16. Let $a_1, a_2, \ldots$ and $b_1, b_2, \ldots$ be arbitrary sequences of complex numbers. Prove Lagrange's formula:

$$\left| \sum_{i=1}^{n} a_i b_i \right|^2 = \left(\sum_{i=1}^{n} |a_i|^2 \right) \left(\sum_{i=1}^{n} |b_i|^2 \right) - \sum_{1 \leq i < j \leq n} |a_i \overline{b_j} - a_j \overline{b_i}|^2 \text{ for every } n \geq 1.$$

Note this formula has Cauchy's inequality as a consequence:

$$\left| \sum_{i=1}^{n} a_i b_i \right|^2 \leq \left(\sum_{i=1}^{n} |a_i|^2 \right) \left(\sum_{i=1}^{n} |b_i|^2 \right). \qquad \diamond$$

An *Egyptian fraction decomposition* of a rational number is a way of writing that number as a sum of fractions with numerator 1 and distinct denominators, and the number of terms in that expression is the length of the decomposition. For example,

$$\frac{97}{126} = \frac{1}{2} + \frac{1}{4} + \frac{1}{51} + \frac{1}{4284}$$
$$= \frac{1}{2} + \frac{1}{7} + \frac{1}{9} + \frac{1}{84} + \frac{1}{252}$$

are Egyptian fraction decompositions of $97/126$ of lengths 4 and 5, respectively.

Problem 2.1.17. Obviously 1 has the Egyptian fraction decomposition $1 = 1/1$ of length 1, and a little thought shows that 1 cannot have an Egyptian fraction decomposition of length 2.

Show that 1 has an Egyptian fraction decomposition of length k for every $k \geq 3$. $\qquad \diamond$

Problem 2.1.18. Let r be a positive rational number with $r \leq 1$. Show that r has an Egyptian fraction decomposition. $\qquad \diamond$

Corollary 2.1.19. *Every positive rational number has infinitely many Egyptian fraction decompositions.*

Proof. First we show that every positive rational number s has an Egyptian fraction decomposition.

If $s \leq 1$ we have shown this. Suppose $s > 1$. Since the harmonic series $\sum_{n=1}^{\infty} \frac{1}{n}$ diverges, there is some k such that $\sum_{n=1}^{k} \frac{1}{n} \leq s < \sum_{n=1}^{k+1} \frac{1}{n}$.

If $\sum_{n=1}^{k} \frac{1}{n} = s$ we are done, as that is an Egyptian fraction decomposition of s. If not, let $r = s - \sum_{n=1}^{k} \frac{1}{n}$. Then $r < 1$ so r has an Egyptian fraction decomposition $r = \sum_{i=1}^{j} \frac{1}{a_i}$. Then $s = \sum_{n=1}^{k} \frac{1}{n} + \sum_{i=1}^{j} \frac{1}{a_i}$. But $r < \frac{1}{k+1}$ so $a_i > k+1$ for $i = 1, \dots, j$, and hence the denominators $1, \dots, k, a_1, \dots, a_j$ in this expression are all distinct, so this is an Egyptian fraction decomposition of s.

Now that we know that s has at least one Egyptian fraction decomposition, we can show that it has infinitely many. Let $s = \sum_{i=1}^{m} \frac{1}{a_i}$ be an Egyptian fraction decomposition of s, and order the terms so that $a_1 < a_2 < \dots < a_m$. We have shown that for any $k \neq 2$, 1 has an Egyptian fraction decomposition of length k. Choose such a decomposition $1 = \sum_{j=1}^{k} \frac{1}{b_j}$. Multiplying by $\frac{1}{a_m}$, we see that $\frac{1}{a_m} = \sum_{j=1}^{k} \frac{1}{a_m b_j}$. Then, substituting, $s = \sum_{i=1}^{m-1} \frac{1}{a_i} + \sum_{j=1}^{k} \frac{1}{a_m b_j}$. Now $b_1, \dots, b_k$ are all distinct, so $a_m b_1, \dots, a_m b_k$ are all distinct, and these are all greater than or equal to a_m, which is the largest of $a_1, \dots, a_m$, so the denominators $a_1, \dots, a_{m-1}, a_m b_1, \dots, a_m b_k$ are all distinct and so this is an Egyptian fraction decomposition of s of length $m + k - 1$. Finally, these all have different lengths, so there are infinitely many of them. □

Problem 2.1.20. Let $S = \{1, 2, \dots, n\}$ be the set of the first n positive integers. Call two subsets T_1 and T_2 of S neighbors if T_2 can be obtained from T_1 by either adding or deleting a single element.

(a) Show that it is possible to list the 2^n subsets of S in order such that any two consecutive subsets are neighbors, and also the last subset and the first subset are neighbors.

(b) Show that it is possible to list the 2^n subsets of S as in part (a), with the list beginning $\emptyset, \{1\}, \{1, 2\}, \dots, \{1, 2, \dots, n\}$.   ◇

Problem 2.1.21. Let $S = (a_0, a_1, a_2, \dots)$ be a strictly increasing sequence of positive integers, i.e., $a_0 < a_1 < a_2 < \dots$. Suppose we want to represent every positive integer as a sum of distinct elements of S. There are two necessary conditions that we can easily see. First, since we want to represent 1, and the smallest integer in S is a_0, we must have $a_0 = 1$. Second, the largest integer we can represent only using $a_1, \dots, a_k$ is $a_1 + \dots + a_k$, and the smallest integer we can represent using a_{k+1} is a_{k+1}, so we must have $a_{k+1} \leq (a_1 + \dots + a_k) + 1$ as otherwise there will be a "gap" and we will not be able to represent $(a_1 + \dots + a_k) + 1$. A priori, there might be some other, more subtle conditions that are necessary in order for us to be able to represent every positive integer as a sum of distinct elements of S, but in fact there are not; these two conditions are also sufficient. Show this.

That is, let $S = (a_0, a_1, a_2, \dots)$ be a strictly increasing sequence of positive integers such that:

(i) $a_0 = 1$; and

(ii) $a_{k+1} \leq (a_1 + \dots + a_k) + 1$ for every k.

(a) Show that every positive integer can be expressed as a sum of distinct elements of S.

For example, consider $S = (1, 2, 4, 8, 16, 25, 36, 49, 64, 81, 100, \ldots)$. Then $85 = 4 + 81$. But also $85 = 8 + 16 + 25 + 36$, and also $85 = 1 + 4 + 16 + 64$. Thus the representation of 85 is not unique.

(b) Let S be as above. Show that every positive integer can be represented in a unique way as a sum of distinct elements of S if and only if S is the sequence $S = (1, 2, 4, 8, 16, 32, 64, 128, 256, \ldots)$ of powers of 2.   ◇

Problem 2.1.22. (a) Let $x = m_1 + \sqrt{m_1^2 - 1}$ for some positive integer m_1. Show that, for every $n \geq 1$, $x^n = m_n + \sqrt{m_n^2 - 1}$ for some positive integer m_n.

(For example, if $x = 2 + \sqrt{3} = 2 + \sqrt{2^2 - 1}$, then $x^2 = 7 + 4\sqrt{3} = 7 + \sqrt{48} = 7 + \sqrt{7^2 - 1}$, $x^3 = 26 + 15\sqrt{3} = 26 + \sqrt{675} = 26 + \sqrt{26^2 - 1}$, ...)

(b) More generally, let $x = m_1 + \sqrt{m_1^2 - N}$ for some positive integer m_1 and some integer N. Show that, for every $n \geq 1$, $x^n = m_n + \sqrt{m_n^2 - N^n}$ for some positive integer m_n.

(For example, if $x = 3 + \sqrt{11} = 3 + \sqrt{3^2 + 2}$, then $x^2 = 20 + 6\sqrt{11} = 20 + \sqrt{396} = 20 + \sqrt{20^2 - 4}$, $x^3 = 126 + 38\sqrt{11} = 126 + \sqrt{15884} = 126 + \sqrt{126^2 + 8}$, ...)   ◇

Note that $1^2 = 1^2$ (trivially), $3^2 + 4^2 = 5^2$ (a Pythagorean triple), and also $9^2 + 22^2 + 54^2 = 59^2$ and $2^2 + 4^2 + 13^2 + 30^2 = 33^2$. So we have examples of sums of 1, 2, 3, and 4 squares that are perfect squares.

Problem 2.1.23. Show that, for every positive integer n, there are n distinct perfect squares whose sum is a perfect square, i.e., show that for every positive integer n these are distinct positive integers $a_1, \ldots, a_n$ such that $a_1^2 + \ldots + a_n^2$ is a perfect square.   ◇

Problem 2.1.24. Show that there is a strictly increasing sequence of positive integers $a_1, a_2, \ldots$ such that, for every positive integer n, the sum $a_1^2 + \ldots + a_n^2$ is a perfect square.   ◇

Of course, this problem implies the previous problem. But it is more restrictive than the previous problem. In that problem, given that $a_1^2 + \ldots + a_n^2$ is a perfect square, you must find distinct positive integers such that $b_1^2 + \ldots + b_{n+1}^2$ is a perfect square, but we need not have $b_1 = a_1, \ldots, b_n = a_n$. (A priori it might be the case that every solution to the first problem is indeed a solution to the second problem, but in fact that is not the case.)

Problem 2.1.25. A characteristic of degree d is a 2-by-d array of numbers

$$\begin{bmatrix} m_1 & m_2 & \ldots & m_d \\ n_1 & n_2 & \ldots & n_d \end{bmatrix}$$

with each m_i and each n_i equal to 0 or 1. The characteristic is even if $m_1 n_1 + m_2 n_2 + \ldots + m_d n_d$ is even and is odd if $m_1 n_1 + m_2 n_2 + \ldots + m_d n_d$ is odd.

Let $e(d)$ be the number of even characteristics of degree d and let $o(d)$ be the number of odd characteristics of degree d. Show that

$$e(d) = 2^{d-1}(2^d + 1) \quad \text{and} \quad o(d) = 2^{d-1}(2^d - 1)$$

for every $d \geq 1$.   ◇

Problem 2.1.26. Fix positive integers r, s, and t and define a sequence by

$$a_1 = r, \qquad a_{n+1} = (s+1)a_n + t \quad \text{for } n \geq 1.$$

(a) Suppose $r = s = t = 1$, so that the sequence is defined by $a_1 = 1$, $a_{n+1} = 2a_n + 1$ for $n \geq 1$. Show that $a_n = 2^n - 1$ for every n.

(b) Suppose $r = s = 1$, so that the sequence is defined by $a_1 = 1$, $a_{n+1} = 2a_n + t$ for $n \geq 1$. Show that $a_n = (t+1)2^n - t$ for every n.

(c) Suppose $r = t = 1$, so that the sequence is defined by $a_1 = 1$, $a_{n+1} = (s+1)a_n + 1$ for $n \geq 1$. Show that $a_n = ((s+1)^n - 1)/s$ for every n.

(d) Suppose $s = t = 1$, so that the sequence is defined by $a_1 = r$, $a_{n+1} = 2a_n + 1$ for $n \geq 1$. Show that $a_n = (r+1)2^n - 1$ for every n.

(e) Discover and prove a formula for a_n in general.   ◇

Problem 2.1.27. Let m be any positive integer; write m as a string of decimal digits. There is a real number x such that, for every $n \geq 1$, the decimal expansion of x^n contains this string.   ◇

Problem 2.1.28. (a) Let S be any set of $n + 1$ distinct integers between 1 and $2n$. Show that S contains two consecutive integers.

(b) Let S be any set of $n + 1$ distinct integers between 1 and $2n$. Show that S contains two integers whose sum is $2n + 1$.   ◇

Problem 2.1.29. (a) Let S be any set of $n + 1$ distinct integers between 1 and $2n$. Show that S contains two distinct integers x and y with x dividing y.

(b) Let S be any set of $n + 1$ distinct odd integers between 1 and $3n$. Show that S contains two distinct integers x and y with x dividing y.   ◇

Problem 2.1.30. Let S be any set of n not necessarily distinct integers.

(a) Show that there is a nonempty subset T of S such that the sum of the elements of T is divisible by n.

(b) Suppose that $n \geq 5$. Show that there is a nonempty subset $U = \{a_1, \ldots, a_k\}$ of S and signs $\varepsilon_1 = \pm 1, \ldots, \varepsilon_k = \pm 1$ such that $\varepsilon_1 a_1 + \ldots + \varepsilon_k a_k$ is divisible by n^2.

(Note that we need $n \geq 5$ in (b) as for $n = 4$, (b) is false for $S = \{1, 2, 4, 8\}$.)
                                                                              ◇

Problem 2.1.31. For a real number x, we let $\{x\} = x - [x]$, where $[x]$ is the largest integer less than or equal to x. Note $0 \leq \{x\} < 1$.

Let x be any real number that is not rational. Let $S = \{ \{nx\} \mid n \text{ an integer } \}$. Show that S is dense in the interval $[0, 1]$. That is, let y be any real number with

$0 \le y \le 1$ and let ε be any positive real number. Show that there is an integer n such that $|\{nx\} - y| < \varepsilon$. ◇

Problem 2.1.32. A lattice point is a point (m, n) in the plane with both coordinates m and n integers.

(a) Let $t(m, n)$ be a function defined on lattice points with the following properties:

(i) $t(m, n)$ is a nonnegative integer for every lattice point (m, n); and

(ii) the value of t at any lattice point is the average of its values at that point's four immediate neighbors, i.e., $t(m, n) = (1/4)(t(m + 1, n) + t(m - 1, n) + t(m, n + 1) + t(m, n - 1))$ for every lattice point (m, n).

Show that $t(m, n)$ must be constant.

(b) Let $t(m, n)$ be a function defined on lattice points with the following properties:

(i) $t(m, n)$ is a nonnegative integer for every lattice point (m, n); and

(ii) the value of t at any lattice point is the average of its values at that point's immediate neighbors to the right and above, i.e., $t(m, n) = (1/2)(t(m + 1, n) + t(m, n + 1))$ for every lattice point (m, n).

Show that $t(m, n)$ must be constant. ◇

Problem 2.1.33. (a) Let P be any polygon in the plane. Show that P can be divided into triangles, all of whose vertices are vertices of P.

(b) A lattice polygon in the plane is a polygon all of whose vertices are lattice points. Prove Pick's theorem:

Theorem 2.1.34. *Let P be a lattice polygon in the plane. Then*

$$\text{Area}(P) = a + b/2 - 1$$

where a is the number of lattice points in the interior of P and b is the number of lattice points on the boundary of P. ◇

By way of clarification of (a), suppose that P is a rectangle. Then P may be divided into two triangles by drawing either of the two diagonals, and that is allowed. But we do not allow a division of P into four triangles by drawing both of the diagonals, as that would introduce a new vertex, at the center of the rectangle, that was not one of the original vertices of P.

As an illustration of Pick's theorem, let P be the quadrilateral whose edges are the line segments from $(0, 0)$ to $(6, 1)$, from $(6, 1)$ to $(4, 2)$, from $(4, 2)$ to $(3, 3)$, and from $(3, 3)$ back to $(0, 0)$. Then there are 5 lattice points in the interior of P, the points $(2, 1)$, $(3, 1)$, $(4, 1)$, $(5, 1)$ and $(3, 2)$, and 6 lattice points on the boundary of P, the points $(0, 0)$, $(6, 1)$, $(4, 2)$, $(3, 3)$, $(2, 2)$ and $(1, 1)$, and P has area $5 + 6/2 - 1 = 7$.

Problem 2.1.35. Suppose that the plane is cut into regions by a finite number of straight lines. Show that the regions can be colored black or white in such a way that no two adjacent regions have the same color. (We consider two regions to be adjacent if they share an edge.)   ◇

Problem 2.1.36. (a) You have coins $C_1, C_2, \ldots$. For each k, the coin C_k has probability $1/(2k-1)$ of landing heads. If the first n coins $C_1, C_2, \ldots, C_n$ are tossed, find, with proof, the probability h_n of obtaining an odd number of heads.

(b) This time, for each k, the coin C_k has probability $k/(2k-1)$ of landing heads. If the first n coins $C_1, C_2, \ldots, C_n$ are tossed, find, with proof, the probability h_n of obtaining an odd number of heads.

(c) Now for the general case: For each k, the coin C_k has probability p_k of landing heads. If the first n coins $C_1, C_2, \ldots, C_n$ are tossed, find, with proof, the probability h_n of obtaining an odd number of heads.   ◇

Your answer to part (c) will have some interesting consequences:

(a) If $p_k = 1/2$ for some value of k, then $h_n = 1/2$ for every $n \geq k$, i.e., once you toss a fair coin, it is equally likely that you will have an odd and an even number of heads from that point on.

(b) If no $p_k = 1/2$, then no $h_n = 1/2$, i.e., if you never toss a fair coin, it will never be the case that it is equally likely that you will have an odd and an even number of heads.

(c) If there is some $\varepsilon > 0$ such that $\varepsilon \leq p_k \leq 1 - \varepsilon$ for every k, then $\lim_{n \to \infty} h_n = 1/2$, i.e., if there is some nontrivial uniform bound on the unfairness of the coins, then once you toss enough coins it will become almost equally likely that you will have an odd and an even number of heads. (It is easy to see you need some restriction. For example, if $h_n \neq 1/2$, and if $p_k = 0$ for $k > n$, i.e., if you never obtain a head after the nth toss, then the probability of an odd number of heads won't change, i.e., $h_{n+1} = h_{n+2} = \ldots$ will all equal h_n. And even if p_k is not exactly 0 for $k > n$, if we choose $p_{n+1}, p_{n+2}, \ldots$ so small that the probability of obtaining at least one head in the tosses after the nth toss is extremely small, then h_n can't possibly change enough to make it approach $1/2$.)

Problem 2.1.37. An urn contains w white balls and b black balls. Balls are drawn at random from the urn until all the remaining balls are of the same color. Find, with proof, the probability that they are all white.   ◇

Problem 2.1.38. A finite *graph* consists of a finite number of vertices, and a finite number of edges, where an edge joins two distinct vertices, and no two vertices may be joined by more than one edge. A finite *pseudograph* is a generalization of a graph, where we allow two vertices to be joined by more than one edge, and we also allow "loops," that is, edges joining a vertex to itself. A directed pseudograph is a pseudograph in which we assign a direction to each edge.

The *valence* of a vertex in a finite pseudograph, or directed pseudograph, is the number of edges incident to that vertex. The *in-valence* (resp. *out-valence*) of a vertex in a directed pseudograph is the number of edges coming into (resp. out of) that vertex.

A *path* in a pseudograph is an unbroken path from one vertex to another (perhaps the same one). In other words, it consists of an edge joining vertex v_0 to vertex v_1, followed by an edge joining vertex v_1 to v_2, ..., finishing with an edge joining vertex v_{n-1} to vertex v_n. We say that such a path, which consists of n edges, has *length n*. In a directed pseudograph, we require that each edge be traversed in the given direction. A pseudograph is connected if there is a path joining any two vertices.

Prove the following.

Theorem 2.1.39. *(a) Let G be a finite connected pseudograph. There is a path containing all the edges of G exactly once if and only if one of the following two conditions is true:*

(i) Every vertex of G has even valence, or

(ii) There are exactly two vertices v_0 and v_1 of G with odd valence.

Furthermore, in case (i) such a path may begin at any vertex and must necessarily end at the same vertex, and in case (ii) such a path must begin at one of the vertices of odd valence, can begin at either of these two vertices, and must necessarily end at the other.

(b) Let G be a finite directed connected pseudograph. There is a path containing all the edges of G exactly once if and only if one of the following two conditions is true:

(i) Every vertex of G has even valence, and for every vertex of G, its in-valence is equal to its out-valence, or

(ii) There are exactly two vertices v_0 and v_1 of G with odd valence. Also, for every even vertex of G, its in-valence is equal to its out-valence, while the out-valence of v_0 is 1 more than its in-valence and the in-valence of v_1 is 1 more than its out-valence.

Furthermore, in case (i) such a path may begin at any vertex and must necessarily end at the same vertex, and in case (ii) such a path must begin at v_0 and must necessarily end at v_1. ◇

Euler originally proved this result for a graph, in solving the famous *Seven Bridges of Königsberg* problem, but the result for a pseudograph, or a directed pseudograph, is no more difficult to prove.

Now we come to some recreational problems.

Problem 2.1.40. (a) Consider an m-by-n checkerboard, with m and n both odd, with the corners colored white. Show that if any white square is removed from the checkerboard, the remaining squares can be tiled by dominoes.

(A domino is a 1-by-2 rectangle that covers exactly two squares of the checkerboard. A tiling is a covering where no two pieces overlap.) Note that the checkerboard has 1 more white square than black square, and that a domino always covers 1 square of each color. Thus, in order for a domino to cover a sub-board, that subboard must have an equal number of white and black squares. Hence the requirement to remove a white square.

(b) Consider an m-by-n checkerboard, with at least one of m and n even. Show that if any two squares of opposite colors are removed from the checkerboard, the remaining squares can be tiled by dominoes, except in the following case: The smaller of m and n is 1, and the leftmost/uppermost removed square is in an even position, or equivalently the rightmost/lowermost removed square is in an odd position (counting from left to right/top to bottom).

Again the requirement that the squares have opposite colors is to ensure that the subboard has an equal number of white and black squares. Note that in the case we have excluded it is not possible to tile the subboard, as in that case this subboard will break up into two or three disjoint subboards, two of which will have an odd number of squares, and hence they will be unable to be tiled by dominoes.   ◇

Problem 2.1.41. Show that if any square is removed from a 2^n-by-2^n checkerboard, the remaining squares can be tiled by L-trominoes.

(An L-tromino is an L-shaped figure that covers exactly three squares of the checkerboard.)   ◇

Problem 2.1.42. Recall we defined an n-by-n antimagic square to be an n-by-n square of numbers such that the sums of each of the rows, each of the columns, and the two diagonals are all distinct.

(a) Show that for every $n \geq 2$, there is an n-by-n antimagic square all of whose entries are positive integers.

(b) Show that for every $n \geq 2$, there is an n-by-n antimagic square all of whose entries are distinct positive integers.   ◇

Problem 2.1.43. It is easy to arrange n rooks on an n-by-n chessboard such that no two rooks attack each other: Just put them all on one of the diagonals. But that arrangement is not very interesting (or challenging). Show that, for every $n \geq 4$, it is possible to arrange n rooks on an n-by-n chessboard such that no two rooks are attacking each other, and no rook is on either (or both) of the two diagonals.   ◇

Problem 2.1.44. In bean solitaire, you begin with $3n$ beans distributed in three piles with a, b, and c beans, respectively. You may make either of the following moves:

(a) You may remove and discard the same number of beans from each pile.

(b) If some pile has an even number of beans, you may transfer half of those beans to one of the other piles.

The object of the game is to empty all the piles. (Note that you remove a multiple of three beans in move (a), while you do not change the number of beans in move (b), so to have any chance of winning the number of beans you start with must be divisible by three.)

Here is a typical game:

5	13	12
11	13	6
6	8	1
6	4	5
2	0	1
1	1	1
0	0	0

Can you always win at bean solitaire? No. For example, if you begin with

3	0	0

there is nothing you can do.

At this point, we do what everyone playing solitaire does in this situation: cheat. By cheating, we mean adding a bean to each pile. Then in this situation we proceed

3	0	0
4	1	1
2	3	1
1	2	0
1	1	1
0	0	0

and we have won.

Thus, to be precise, the rules for bean solitaire with cheating are as follows:

You begin with $3n + 3$ beans, with $3n$ beans distributed in piles of a, b, and c beans, respectively, and the remaining 3 beans held in reserve.

You may make any of the following moves:

(a) You may remove the same number of beans from each pile and transfer them to the reserve.

(b) If some pile has an even number of beans, you may transfer half of those beans to one of the other piles.

(c) You may take three beans from the reserve and add one of them to each of the piles.

You win if you empty all the piles.

Show that it is always possible to win at bean solitaire with cheating.   ◇

2.2 Problems and Results in Elementary Number Theory

We have already seen how to prove one of the basic theorems of elementary number theory, the division algorithm, by induction. In this section, we will develop a good bit of number theory, with some key results to be proved by induction, and others following from them. We also include a few results that are less basic, but nevertheless important.

Problem 2.2.1. Let a and b be any two nonnegative integers, not both 0. Show that there is a unique positive integer d such that:

(a) d divides both a and b, and

(b) if e is any integer that divides both a and b, then e divides d.   ◇

This integer d is called the *greatest common divisor* of a and b, $d = \gcd(a, b)$. It is certainly the largest common divisor, but that is not what is important. What is important is that every common divisor divides it. Thus it might be better to call it the most divisible common divisor rather than the greatest common divisor, but the term greatest common divisor has been in use for millennia, so we will continue to use it.

Problem 2.2.2. Let a and b be any two nonnegative integers, not both 0, and let $d = \gcd(a, b)$. Show there are integers x and y such that

$$d = ax + by.$$

(In this situation d is said to be an integer linear combination of a and b.)   ◇

(The preceding two problems can be done separately, but it is even easier to do them simultaneously.)

Corollary 2.2.3. *Let a and b be any two nonnegative integers, not both 0. An integer n can be expressed as $n = au + bv$ for some integers u and v if and only if d divides n.*

Proof. Suppose that $n = au + bv$. Since d divides a, it divides au, and since d divides b, it divides bv; then d divides their sum $au + bv = n$. Conversely, suppose d divides n; write $n = dn'$. Then $d = ax + by$ so $n = dn' = (ax + by)n' = a(xn') + b(yn')$.   □

Definition 2.2.4. Let a and b be any two nonnegative integers, not both 0. Then a and b are *relatively prime* if $\gcd(a, b) = 1$.

A set of nonnegative integers $\{a_1, \ldots, a_k\}$ is *pairwise relatively prime* if any pair a_i and a_j, $i \neq j$, is relatively prime.

Here is a *very* important consequence of this problem.

Theorem 2.2.5 (Euclid's Lemma). *Let m and n be any two positive integers. Let a be any integer that divides the product mn. If a and m are relatively prime, then a divides n.*

Proof. Since a and m are relatively prime, we have that $1 = ax + my$ for some integers x and y. Then $n = n1 = n(ax + my) = a(nx) + (mn)y$. Now a certainly divides the first term $a(nx)$, and a divides the second term $(mn)y$, since, by hypothesis, a divides mn; hence a divides their sum, which is n.   □

Here are some consequences of Euclid's lemma.

Corollary 2.2.6. *Let a and b be nonzero relatively prime positive integers. Let c be a positive integer and suppose that a divides c and that b divides c. Then ab divides c.*

Proof. Since a divides c, we may write $c = ae$ for some e. Then b divides $c = ae$ and a and b are relatively prime, so by Euclid's lemma, b divides e, i.e., $e = bf$ for some f. But then $c = ae = a(bf) = (ab)f$ and so ab divides c.   □

Corollary 2.2.7. *Let a, b, and c be positive integers. Suppose that a and b are relatively prime and that a and c are relatively prime. Then a and bc are relatively prime.*

Proof. Let $d = \gcd(a, bc)$ and let $e = \gcd(d, b)$. Then e divides d, and d divides a, so e divides a. Also, e divides b. Thus, e divides $\gcd(a, b) = 1$, so $e = 1$, i.e., d and b are relatively prime.

Now d divides bc, and d and b are relatively prime, so, by Euclid's lemma, d divides c. But d divides a, so d is a common divisor of a and c. But a and c are relatively prime, so $d = 1$, i.e., a and bc are relatively prime.   □

Corollary 2.2.8. *(1) Let a and b be any two nonnegative integers, not both 0. Let c be any positive integer. Then*

$$\gcd(ca, cb) = c\gcd(a, b).$$

(2) Let a and b be any two nonnegative integers, not both 0. Let c be a common divisor of a and b. Then

$$\gcd(a/c, b/c) = \gcd(a, b)/c.$$

In particular, a/c and b/c are relatively prime if and only if $c = \gcd(a, b)$.

Proof. (1) Let $d = \gcd(a, b)$ and $e = \gcd(ca, cb)$. Then $a = da'$ for some integer a', in which case $ca = cda'$. Similarly, $b = db'$ for some integer b', in which case $cb = cdb'$. Thus cd is a common divisor of ca and cb, so cd divides e.

Now $d = ax + by$ for some integers x and y, so $cd = c(ax + by) = (ca)x + (cb)y$ is an integer linear combination of ca and cb, so e divides cd by a previous corollary.

Hence $e = cd$.

(2) Write $a = ca'$ and $b = cb'$. Then, of course, $a' = a/c$ and $b' = b/c$. Now, by part (1), $\gcd(a, b) = \gcd(ca', cb') = c\gcd(a', b')$. But then we immediately see that $\gcd(a, b)/c = \gcd(a', b') = \gcd(a/c, b/c)$.

If we let $d = \gcd(a, b)$, then this equation becomes $\gcd(a/c, b/c) = d/c$, so a/c and b/c are relatively prime, i.e., $\gcd(a/c, b/c) = 1$, if and only if $d/c = 1$, i.e., if and only if $c = d = \gcd(a, b)$. $\qquad\qquad\square$

Note this whole chain of results was a consequence of the single induction argument above.

We recall some standard terminology. The number 1 is a *unit*. An integer $p > 1$ that is only divisible by 1 and p is a *prime*. A positive integer that is neither 1 nor a prime is *composite*.

Corollary 2.2.9. *Let p be a prime and suppose that p divides the product ab of two positive integers a and b. Then p divides a or p divides b.*

Proof. Let g be the gcd of p and a. Then, in particular, g divides p, so $g = p$ or $g = 1$. If $g = p$, then, since g divides a, we immediately have that p divides a. If $g = 1$, then p and a are relatively prime. But we are assuming that p divides ab, so in this case we conclude from Euclid's lemma that p divides b. $\qquad\square$

Problem 2.2.10. (a) Let $\{a_1, a_2, \ldots, a_k\}$ be a set of pairwise relatively prime positive integers. Suppose that a_i divides the positive integer b for each $i = 1, \ldots, k$. Show that the product $a_1 a_2 \cdots a_k$ divides b.

(b) Let $\{b_1, b_2, \ldots, b_k\}$ be a set of positive integers. Suppose that a is a positive integer with a and b_i relatively prime for each $i = 1, \ldots, k$. Show that a and the product $b_1 b_2 \cdots b_k$ are relatively prime.

(c) Let p be a prime and let $\{c_1, c_2, \ldots, c_k\}$ be a set of positive integers. Suppose that p divides the product $c_1 c_2 \cdots c_k$. Show that p divides c_i for some i.

$\diamond$

The following theorem is of the highest level of importance. Indeed, as its name implies, it is fundamental. (In this theorem, we regard 1 as having the empty factorization.)

Problem 2.2.11. Prove the following theorem:

Theorem 2.2.12 (The fundamental theorem of arithmetic). *Let n be any positive integer. Then n has a factorization into primes,*

$$n = p_1^{e_1} p_2^{e_2} \cdots p_k^{e_k} \quad \text{with } p_1, p_2, \ldots, p_k \text{ distinct primes}$$
$$\text{and } e_1, e_2, \ldots, e_k \text{ positive integers,}$$

and this factorization is unique up to the order of the factors. $\qquad\qquad\diamond$

Here is another important consequence:

Theorem 2.2.13. *Let n be a positive integer, n ≠ 1. The following are equivalent:*

(a) *If n divides a product st of two positive integers s and t, then n divides s or n divides t.*

(b) *If m is a positive integer dividing n, then m = 1 or m = n.*

Proof. First suppose (a) is true. We want to show (b) is true. So let m be a positive integer dividing n. Then $n = mq$ for some positive integer q, in which case q also divides n. Since (a) is true, n must divide m or q. If n divides m, then each of m and n divide each other, so they are equal, $m = n$. If q divides m, then each of q and n divide each other, so they are equal, $q = n$, and then $m = 1$.

Now suppose (b) is true. We want to show (a) is true. So suppose that n divides st. Let $d = \gcd(n, s)$. Then, in particular, d divides n, so by (b), $d = 1$ or $d = n$. If $d = n$, then, as d divides s, we have that n divides s. If $d = 1$, then, by definition, n and s are relatively prime. Hence in this case, by Euclid's lemma, n divides t. Hence, in any event, n divides s or t.   □

Without going into the details, we can consider the analogues of these conditions in more general algebraic systems. The analog of (a) is the definition of a *prime* while the analog of (b) is the definition of an *irreducible*. Of course, for the positive integers, it is (b) that is precisely the definition of a prime. Thus these two conditions for the positive integers are the "reverse" of the conditions in more general situations. But this lemma tells us that's OK; in the positive integers these two conditions are equivalent, so it does not matter which one we use as the definition of a prime. (In general, these two conditions are not equivalent, but that is something we cannot go into in detail here. But we will observe that the proof that (a) implies (b) was perfectly general, and that continues to hold. The proof that (b) implies (a), however, used Euclid's lemma in a key way, and that we derived as a consequence of induction, so we only have it for the positive integers. It is true somewhat more generally, but not in complete generality, and (b) does not imply (a) in complete generality.)

Here is another consequence of our inductive arguments.

Theorem 2.2.14 (The Chinese remainder theorem). *Let $\{a_1, \ldots, a_k\}$ be pairwise relatively prime and let $\{b_1, \ldots, b_k\}$ be arbitrary integers. Then the system of simultaneous congruences*

$$x \equiv b_1 \ (\mathrm{mod}\ a_1)$$
$$x \equiv b_2 \ (\mathrm{mod}\ a_2)$$
$$\vdots$$
$$x \equiv b_k \ (\mathrm{mod}\ a_k)$$

has a solution, and that solution is unique mod $a_1 a_2 \cdots a_k$.

Proof. First we show existence, then we show uniqueness.

For each i, let z_i be the product of all of the integers $a_1, a_2, \ldots, a_k$ *except* a_i. Since, by assumption, a_i is relatively prime to each a_j, $j \neq i$, a_i is relatively prime to their product z_i. Thus there are integers x_i and y_i with $1 = a_i x_i + z_i y_i$. Let $s_i = z_i y_i b_i$. We claim that

$$s_i \equiv b_i \ (\mathrm{mod} \ a_i) \quad \text{and} \quad s_i \equiv 0 \ (\mathrm{mod} \ a_j) \text{ for each } j \neq i.$$

To see the first congruence, note that, multiplying the above equation by b_i, we have that $b_i = a_i x_i b_i + z_i y_i b_i = a_i x_i b_i + s_i \equiv s_i \ (\mathrm{mod} \ a_i)$ as $a_i x_i b_i$ is a multiple of a_i. To see the second congruence, note that $s_i = z_i y_i b_i \equiv 0 \ (\mathrm{mod} \ a_j)$ for each $j \neq i$, as y_i is a multiple of a_j.

Now let

$$s = s_1 + s_2 + \ldots + s_k.$$

Then

$$s \equiv 0 + \ldots + 0 + b_i + 0 + \ldots + 0 \equiv b_i \ (\mathrm{mod} \ a_i)$$

for each i, and so s is a solution of this system of simultaneous congruences.

Now if $s' \equiv s \ (\mathrm{mod} \ a_1 a_2 \cdots a_k)$, then $s' \equiv s \ (\mathrm{mod} \ a_i)$ for each i, so s' is also a solution.

We claim that these are all the solutions. To see this, consider an arbitrary solution t. Then $t \equiv s \ (\mathrm{mod} \ a_i)$ for each i, so $t - s \equiv 0 \ (\mathrm{mod} \ a_i)$ for each i, i.e., $t - s$ is divisible by a_i for each i. Since $a_1, a_2, \ldots, a_k$ are pairwise relatively prime, that implies $t - s$ is divisible by the product $a_1 a_2 \cdots a_k$, i.e., $t - s \equiv 0 \ (\mathrm{mod} \ a_1 a_2 \cdots a_k)$, and so $t \equiv s \ (\mathrm{mod} \ a_1 a_2 \cdots a_k)$. □

These are all important theoretical results, but we have not yet addressed the problem of how to effectively do computations. Of course, for small integers, we can use trial and error. But now we develop an extremely effective computational method.

Problem 2.2.15. Let a and b be nonnegative integers, not both 0. *Euclid's algorithm* is the following algorithm to find $g = \gcd(a, b)$:

(a) Set $a_0 = a$ and $a_1 = b$.

(b) Given a_n and a_{n+1}, there are two possibilities:

  (i) $a_{n+1} \neq 0$: Let a_{n+2} be the unique nonnegative integer defined by $a_n = a_{n+1} q_n + a_{n+2}$ where q_n is an integer and a_{n+2} is an integer with $0 \leq a_{n+2} < a_{n+1}$. Increment n and loop.

  (ii) $a_{n+1} = 0$: Set $g = a_n$ and stop.

Show that this algorithm is correct, i.e., show that Euclid's algorithm always yields $\gcd(a, b)$.   ◇

Example 2.2.16. (a) Here is an example of Euclid's algorithm in action. Suppose we want to find $\gcd(654321, 123456)$. We have:

$$654321 = 123456 \cdot 5 + 37041$$
$$123456 = 37041 \cdot 3 + 12333$$
$$37041 = 12333 \cdot 3 + 42$$
$$12333 = 42 \cdot 293 + 27$$
$$42 = 27 \cdot 1 + 15$$
$$27 = 15 \cdot 1 + 12$$
$$15 = 12 \cdot 1 + 3$$
$$12 = 3 \cdot 4$$

and so we conclude $\gcd(654321, 123456) = 3$.

(b) Now let us use this computation to find integers x and y with $3 = 654321x + 123456y$. We do this by solving from the bottom up, beginning with the next to the last equation, which has the gcd 3 as its last term:

$$3 = 15 + 12(-1)$$
$$= 15 + (27 + (15)(-1))(-1) = 27(-1) + 15(2)$$
$$= 27(-1) + (42 + 27(-1))(2) = 42(2) + 27(-3)$$
$$= 42(2) + (12333 + 42(-293))(-3) = 12333(-3) + 42(881)$$
$$= 12333(-3) + (37041 + 12333(-3))(881) = 37041(881) + 12333(-2646)$$
$$= 37041(881) + (123456 + 37041(-3))(-2646) = 123456(-2646)$$
$$+ 37041(8819)$$
$$= 123456(-2646) + (654321 + 123456(-5))(8819)$$
$$= 654321(8819) + 123456(-46741)$$

(c) Now let us solve a system of simultaneous congruences by using this. Consider the system

$$x \equiv 6 \ (\text{mod } 11)$$
$$x \equiv 7 \ (\text{mod } 13)$$
$$x \equiv 8 \ (\text{mod } 15)$$
$$x \equiv 9 \ (\text{mod } 16)$$

Then $a_1 = 11$ and $z_1 = 13 \cdot 15 \cdot 16 = 3120$. We now find x_1 and y_1. By Euclid's algorithm we have:

$$3120 = 11 \cdot 283 + 7$$
$$11 = 7 \cdot 1 + 4$$
$$7 = 4 \cdot 1 + 3$$
$$4 = 3 \cdot 1 + 1$$
$$3 = 1 \cdot 3$$

and then working from the bottom up

$$
\begin{aligned}
1 &= 4 + 3(-1) \\
&= 4 + (7 + 4(-1))(-1) = 7(-1) + 4(2) \\
&= 7(-1) + (11 + 7(-1))(2) = 11(2) + 7(-3) \\
&= 11(2) + (3120 + 11(-283))(-3) = 3120(-3) + 11(851)
\end{aligned}
$$

so $x_1 = 851$ and $y_1 = -3$. Then

$$s_1 = z_1 y_1 b_1 = (3120)(-3)(6) = -56160.$$

Similarly we find x_2, y_2, and s_2; x_3, y_3, and s_3; x_4, y_4, and s_4. We skip the details and simply give the answers:

$$
\begin{array}{lll}
x_2 = -203, & y_2 = 1, & s_2 = 18480; \\
x_3 = -305, & y_3 = 2, & s_3 = 36608; \\
x_4 = -134, & y_4 = 1, & s_4 = 19305.
\end{array}
$$

Then the solution to our congruences is

$$s \equiv s_1 + s_2 + s_3 + s_4 = -56160 + 18480 + 36608 + 19305 = 18233 \ (\text{mod } 34320).$$

Now we return to some other important theoretical results.

Problem 2.2.17. Prove the following result:

Theorem 2.2.18 (Fermat's little theorem). *Let p be a prime and let a be any integer. Then*

$$a^p \equiv a \ (\text{mod } p). \qquad \qquad \diamond$$

Note that this theorem states that if p is prime, then for any integer a, p divides $a^p - a = a(a^{p-1} - 1)$. If p divides a this is trivial. If p does not divide a then, since p is prime, p and a are relatively prime, and so, by Euclid's lemma, we have in this case that p divides $a^{p-1} - 1$. Thus Fermat's little theorem is often just stated in the nontrivial case: *If p does not divide a, then $a^{p-1} \equiv 1 \ (\text{mod } p)$.*

Problem 2.2.19. The *Euler totient function* $\varphi(n)$ is defined by:

$\varphi(n) = $ the number of integers between 1 and n that are relatively prime to n.

Prove the following result.

Theorem 2.2.20 (Euler). *Let n be an arbitrary positive integer and let a be any integer that is relatively prime to n. Then*

$$a^{\varphi(n)} \equiv 1 \ (\text{mod } n). \qquad \qquad \diamond$$

Observe that if $n = p$ is a prime, then every integer k with $1 \leq k \leq n - 1$ is relatively prime to n, so that $\varphi(p) = p - 1$ for every prime p. Thus we see that this theorem of Euler is a generalization of Fermat's little theorem.

Problem 2.2.21. Let $\varphi(n)$ be the Euler totient function, as in the last problem. Of course, $\varphi(1) = 1$. Let $n > 1$ be an integer and suppose that n has prime factorization $n = p_1^{e_1} p_2^{e_2} \cdots p_k^{e_k}$. Show that

$$\varphi(n) = (p_1 - 1)p_1^{e_1 - 1}(p_2 - 1)p_2^{e_2 - 1} \cdots (p_k - 1)p_k^{e_k - 1}. \qquad \diamond$$

Problem 2.2.22. Let n be any positive integer. If a is any integer with $a \equiv 1 \pmod{n}$, show that $a^{n^i} \equiv 1 \pmod{n^{i+1}}$ for every positive integer i. $\qquad \diamond$

Problem 2.2.23. Let n be a positive integer and let $n = b_1 b_2 \cdots b_k$ be any factorization of n. Let m be any positive integer dividing n. Show that m has a factorization $m = a_1 a_2 \cdots a_k$ with a_i dividing b_i for each $i = 1, \ldots, k$. $\qquad \diamond$

Problem 2.2.24. Let n be a positive integer and let $n = p_1^{e_1} p_2^{e_2} \cdots p_k^{e_k}$ be the prime factorization of n. Let $r = m/n$ be any rational number, where m is an integer. Show that there are unique integers $a_0, a_{1,1}, \ldots, a_{1,e_1}, a_{2,1}, \ldots, a_{2,e_2}, \ldots, a_{k,1}, \ldots, a_{k,e_k}$ with $0 \leq a_{i,j} < p_i$, $j = 1, \ldots, e_i$, for each $i = 1, \ldots, k$, such that

$$
\begin{aligned}
r &= \frac{m}{n} \\
&= a_0 + \frac{a_{1,1}}{p_1} + \cdots + \frac{a_{1,e_1}}{p_1^{e_1}} + \frac{a_{2,1}}{p_1} + \cdots + \frac{a_{2,e_2}}{p_2^{e_2}} + \cdots + \frac{a_{k,1}}{p_k} + \cdots + \frac{a_{k,e_k}}{p_k^{e_k}}.
\end{aligned}
$$

For example,

$$\frac{511}{600} = -1 + \frac{1}{2} + \frac{0}{4} + \frac{1}{8} + \frac{2}{3} + \frac{2}{5} + \frac{4}{25}. \qquad \diamond$$

(This way of decomposing the rational number m/n is an analog of the familiar method of partial fractions for decomposing quotients of polynomials.)

2.3   Binomial Coefficients and Related Matters

We now define the binomial coefficients. There are several approaches to doing so, and depending on the approach, what is a definition and what is a theorem changes. We adopt the approach where the binomial coefficients are defined recursively.

Consider the following array, known as *Pascal's triangle*:

$$
\begin{array}{ccccccccccccc}
 & & & & & & 1 & & & & & & \\
 & & & & & 1 & & 1 & & & & & \\
 & & & & 1 & & 2 & & 1 & & & & \\
 & & & 1 & & 3 & & 3 & & 1 & & & \\
 & & 1 & & 4 & & 6 & & 4 & & 1 & & \\
 & 1 & & 5 & & 10 & & 10 & & 5 & & 1 & \\
1 & & 6 & & 15 & & 20 & & 15 & & 6 & & 1 \\
\end{array}
$$

In this triangle, each entry is the sum of the entries immediately to its left and right in the row above it. (We consider that there is an infinite sea of 0 entries outside the confines of the triangle.)

We number the rows beginning with the top row being row 0, and we number the northeast-southwest diagonals with the top left diagonal being 0. We denote the entry in the (n, k) position (i.e., in row n and diagonal k) by $\binom{n}{k}$. Thus, with this notation, the entries in Pascal's triangle are given by the recursion

$$
\binom{n+1}{k} = \binom{n}{k-1} + \binom{n}{k} \text{ where } \binom{0}{0} = 1 \text{ and } \binom{0}{k} = 0 \text{ for } k \neq 0.
$$

Problem 2.3.1. Show that

$$
\binom{n}{k} = \binom{n}{n-k} \text{ for every } n \text{ and } k.
$$

◊

Problem 2.3.2. Show that

$$
\sum_{k=0}^{n} \binom{n}{k} = 2^n \text{ for every } n \geq 0
$$

and

$$
\sum_{k \text{ odd}} \binom{n}{k} = \sum_{k \text{ even}} \binom{n}{k} = 2^{n-1} \text{ for every } n \geq 1.
$$

◊

Problem 2.3.3. Show that, for every nonnegative integer n,

Theorem 2.3.4 (The binomial theorem).

$$
(x+y)^n = \sum_{k=0}^{n} \binom{n}{k} x^{n-k} y^k.
$$

◊

For this reason, the numbers $\binom{n}{k}$ are called *binomial coefficients*.

Problem 2.3.5. Show that, for every nonnegative integer n and every integer k,

$$\binom{n}{k} = \text{the number of } k\text{-element subsets of an } n\text{-element set.}$$ ◇

The binomial theorem gives another proof of an earlier result.

Corollary 2.3.6.

$$\sum_{k=0}^{n} \binom{n}{k} = 2^n \text{ for every } n \geq 0$$

and

$$\sum_{k \text{ odd}} \binom{n}{k} = \sum_{k \text{ even}} \binom{n}{k} = 2^{n-1} \text{ for every } n \geq 1.$$

Proof. The first statement follows by setting $x = 1$ and $y = 1$ in the binomial theorem while the second statement follows by setting $x = 1$ and $y = -1$ in the binomial theorem. □

Note that we obtain a k-element subset of an n-element set by choosing any k out of the n elements, where we are not allowed to choose the same element more than once, and the order in which we choose the elements does not matter. Thus the binomial coefficient $\binom{n}{k}$ is usually expressed as "n choose k."

This interpretation of binomial coefficients gives another (indeed, a more conceptual) proof of an earlier result.

Corollary 2.3.7.

$$\binom{n}{k} = \binom{n}{n-k} \text{ for every } n \text{ and } k.$$

Proof. The number of ways of choosing k out of n objects (to keep) is the same as the number of ways of choosing $n - k$ out of n objects (to discard). □

Problem 2.3.8. Show that, for every nonnegative integer n and every integer k with $0 \leq k \leq n$,

$$\binom{n}{k} = \frac{n!}{k!(n-k)!}.$$ ◇

This gives us a more perspicacious proof of an earlier result.

Corollary 2.3.9. *The product of any k consecutive positive integers is divisible by $k!$.*

Proof. We may write any such product P as $P = (j+1)(j+2)\cdots(j+k)$ for some integer $j \geq 0$. But then

$$P = (j+1)(j+2)\cdots(j+k) = \frac{1\cdot2\cdots j(j+1)(j+2)\cdots(j+k)}{1\cdot2\cdots j}$$

$$= \frac{(j+k)!}{j!}.$$

But then

$$\frac{P}{k!} = \frac{(j+k)!}{j!k!} = \binom{j+k}{j}$$

is an integer, and so P is divisible by $k!$.   □

Problem 2.3.10. Show that

$$\sum_{i=0}^{n} \binom{i+k}{k} = \binom{n+k+1}{k+1} \text{ for all nonnegative integers } n \text{ and } k. \quad \diamond$$

Problem 2.3.11. Show that

$$\sum_{n=0}^{\infty} \frac{\binom{n+k+1}{n}}{2^n} = 2^{k+1} \text{ for every nonnegative integer } k. \quad \diamond$$

Problem 2.3.12. Show that, for every integer $n \geq 0$ and every integer i with $0 \leq i \leq n$,

$$\sum_{k=i}^{n} (-1)^k (2k+1)\binom{k+i}{2i} = (-1)^n (n+i+1)\binom{n+i}{2i}. \quad \diamond$$

Problem 2.3.13. For pairs of positive integers (n, k), define $f(n, k)$ by $f(1, k) = 1$ for every k, and

$$f(n, k) = 1 + \sum_{j=1}^{k} f(n-1, j) \text{ for } n > 1.$$

Show that

$$f(n, k) = \binom{n+k-1}{n-1} = \binom{n+k-1}{k}. \quad \diamond$$

We now consider some sampling problems.

Consider an urn containing n distinct balls. We choose k balls from the urn and tabulate the results. In doing so, we note there are two (independent) alternatives as to how to do this:

(a) We may sample with replacement (i.e., after choosing a ball, we note which it is, and then return the ball to the urn before choosing the next ball), or without replacement (i.e., after choosing a ball, we note which it is, and set it aside before choosing the next ball). Thus, in sampling with replacement we may choose the same ball more than once, while in sampling without replacement we cannot do so.

(b) We may decide that order matters (i.e., we consider that for two choices to be the same, we must have drawn the same collection of balls in the same order) or that order is irrelevant (i.e., we consider that two choices are the same if we have drawn the same collection of balls each time, even though we may have drawn them in a different order).

We now wish to count the number of possible outcomes under each of these alternatives.

Theorem 2.3.14. *The following table, counting the results of sampling k out of n objects, is correct:*

	order matters	order irrelevant
without replacement	$P_{n,k} = \dfrac{n!}{(n-k)!}$	$C_{n,k} = \dbinom{n}{k} = \dbinom{n}{n-k}$
with replacement	n^k	$\dbinom{n+k-1}{n-1} = \dbinom{n+k-1}{k}$

except that the entry in the lower right-hand corner is valid for $n + k > 0$; if $n = k = 0$ it should be replaced by 1.

Before beginning the proof we make several observations:
First we note that we have an alternate formula for $P_{n,k}$:

$$P_{n,k} = \frac{n!}{(n-k)!} = \frac{n(n-1)\cdots 2 \cdot 1}{(n-k)(n-k-1)\cdots 2 \cdot 1} = n(n-1)\cdots(n-k+1)$$

We also note that $P_{n,n} = n!$. Furthermore, we note that $P_{n,n}$ counts the number of ways of choosing n distinct objects out of a total of n objects, in order. But that simply means we have chosen all of the objects, so $P_{n,n}$ counts the number of ways of ordering n objects.

We call $P_{n,k}$ the number of *permutations* of n objects taken k at a time; in case $k = n$ we simply call this the number of permutations of n objects. We call $C_{n,k}$ the number of *combinations* of n objects taken k at a time.

Proof. First we note that the table is correct for $k = 0$ and any n, when all the entries are equal to 1: If we are choosing 0 balls, under any alternative, there is only one way to do it: Choose nothing.

We also note that the table is correct for $k = 1$ and any n, when all the entries are equal to n: If we are choosing 1 ball, under any alternative, out of n balls, there are exactly n ways to do it: Choose any one of the balls. (This is also true for $n = 0$, as then it is impossible to make a choice, so there are no ways to do it.)

We now proceed by induction on k.

First, the lower left-hand corner. We know the entry is correct for $k = 1$ (and any n). Now suppose the entry is correct for $k - 1$ (and any n). We wish to choose k objects, in order, with replacement. We do this by first choosing $k - 1$ objects, and there are, by the inductive hypothesis, n^{k-1} ways to do this. For each choice, there are n ways to extend it to a choice of k objects, as we may choose any one of the n objects. Thus the number of ways to choose k objects under these conditions is $n^{k-1}n = n^k$, as claimed.

Next, the upper left-hand corner. We know the entry is correct for $k = 1$ (and any n). Now suppose the entry is correct for $k - 1$ (and any n). We wish to choose k objects, in order, with replacement. We do this by first choosing $k - 1$ objects, and there are, by the inductive hypothesis, $P_{n,k-1}$ ways to do this. But here, for each choice, there are $n - (k - 1) = n - k + 1$ ways to extend it to a choice of k objects, as we are choosing without replacement. (We have already used $k - 1$ of the n objects, and we cannot reuse any of them, so we have $n - (k - 1)$ objects left to choose from.) Thus the number of ways to choose k objects under these conditions is $P_{n,k-1}(n - k + 1)$, as claimed. $(P_{n,k-1}(n - k + 1) = (n(n - 1)\cdots(n - (k - 1) + 1))(n - (k - 1)) = n(n - 1)\cdots(n - k + 2)(n - k + 1).)$

Now for the upper right-hand corner. Actually, we have already seen this result, so this proof is just a new perspective on it. Consider $P_{n,k}$, the number of ways of choosing k out of n objects without replacement, where the order matters. We may consider such a choice as having been done in two steps: First choose the k elements, without caring about the order, and then, for each such choice, decide how to order these chosen elements. The number of ways of doing the first step is, by definition, $C_{n,k}$. The number of ways of doing the second step is the number $P_{k,k}$ of permutations of k objects. Thus we see that $P_{n,k} = C_{n,k}P_{k,k}$. Hence, using our previous results, $C_{n,k} = P_{n,k}/P_{k,k} = (n!/(n - k)!)/k! = n!/(k!(n - k)!) = \binom{n}{k}$.

Finally, we deal with the lower right-hand corner. Here we proceed by induction on n. Let us denote the entry in that corner for a particular value of n and k by $f(n, k)$. To start the induction, we observe that $f(1, k) = 1$ for every k, as if we only have a single object, there is only 1 way to sample: Choose that object k times.

Now suppose we know $f(n - 1, k)$ for every k. Consider a sample of k of n objects. It may be that this sample consists of the nth object, chosen k times. That is 1 possibility. Otherwise, we have chosen the first $n - 1$ objects a total of j times, for some j between 1 and k, in which case we are forced to choose the last object $k - j$ times to obtain a total of k objects. There are $f(n - 1, j)$ ways of doing this. Thus we obtain the recursion

$$f(n, k) = 1 + \sum_{j=1}^{k} f(n - 1, j) \text{ for } n > 1.$$

But we have already shown that the solution to this recursion is $f(n, k) = \binom{n+k-1}{n-1} = \binom{n+k-1}{k}$, as claimed.

Here is a second, more clever proof of this case. In this situation (with replacement, order irrelevant) a sampling is determined by how many times we have chosen each object. We encode this choice as follows: We begin with a straight line with $n + k - 1$ positions, and with k stars and $n - 1$ vertical bars, which we arrange in these positions. If there are j_1 stars to the left of the first vertical bar, j_2 stars between the first and second vertical bar, j_3 stars between the second and third vertical bar, ..., j_{n-1} stars between the $(n-2)$nd and $(n-1)$st (i.e., last) vertical bar, and j_n stars to the right of the last vertical bar, (with $j_1 + j_2 + \ldots + j_n = k$, of course), then we have chosen the first object j_1 times, the second object j_2 times, ..., the last object j_n times. For example, if $n = 4$ and $k = 7$,

$\underline{* *} \mid \underline{* * *} \mid \underline{*} \mid \underline{*}$	encodes	$11, 222, 3, 4$
$\underline{* * *} \mid \underline{* *} \mid \mid \underline{* *}$	encodes	$111, 22, 44$
$\underline{* * * *} \mid \underline{*} \mid \underline{* *} \mid$	encodes	$1111, 2, 33$
$\mid \mid \underline{* * * *} \mid \underline{* * *}$	encodes	$3333, 444$

Now a coding is determined by which $n - 1$ positions out of the total of $n + k - 1$ positions we choose to put the vertical bars in, and the number of these choices is just $C_{n+k-1,n-1} = \binom{n+k-1}{n-1} = \binom{n+k-1}{k}$.   □

Problem 2.3.15. Let p be a prime and let n and m be arbitrary nonnegative integers. Write $n = \sum_{i=0}^k b_k p^k$ with $0 \le b_i < p$ for each i, and $m = \sum_{i=0}^k a_k p^k$ with $0 \le a_i < p$ for each i, so that $b_k b_{k-1} \cdots b_0$ is the base p expansion of n and $a_k a_{k-1} \cdots a_0$ is the base p expansion of m ("padded" with 0's on the left, if necessary, so that the two expansions have the same length). Show that

$$\binom{n}{m} \equiv \binom{b_k}{a_k}\binom{b_{k-1}}{a_{k-1}} \cdots \binom{b_0}{a_0} \pmod{p}.$$ ◊

Example 2.3.16. Let $p = 7$. We easily compute that $480 = 1254_7$, $72 = 132_7$, and $90 = 156_7$. Then

$$\binom{480}{72} = \binom{1254_7}{132_7} \equiv \binom{1}{0}\binom{2}{1}\binom{5}{3}\binom{4}{2} = 1 \cdot 2 \cdot 10 \cdot 6 = 120 \equiv 1 \pmod 7, \text{ and}$$

$$\binom{480}{90} = \binom{1254_7}{156_7} \equiv \binom{1}{0}\binom{2}{1}\binom{5}{5}\binom{4}{6} \equiv 0 \pmod 7 \text{ as } \binom{4}{6} = 0.$$

Problem 2.3.17. Consider the following variant of Pascal's triangle. For a fixed integer a, define $\binom{n}{k}_a$ by

$$\binom{0}{0}_a = 1, \text{ and } \binom{0}{k}_a = 0 \text{ for } k \ne 0;$$

$$\binom{n+1}{k}_a = a\binom{n}{k-1}_a + \binom{n}{k}_a \text{ for } n \ge 0.$$

Find and prove a formula for $\binom{n}{k}_a$.   ◊

Problem 2.3.18. Consider the following variant of Pascal's triangle. For a fixed integer b, define $\binom{n}{k}^b$ by

$$\binom{1}{0}^b = 1, \quad \binom{1}{1}^b = b, \quad \text{and} \quad \binom{1}{k}^b = 0 \text{ for } k \neq 0, 1;$$

$$\binom{n+1}{k}^b = \binom{n}{k-1}^b + \binom{n}{k}^b \quad \text{for } n \geq 1.$$

(Note that $\binom{0}{k}^b$ is not defined.) Find and prove a formula for $\binom{n}{k}^b$. ◇

Problem 2.3.19. Consider the following variant of Pascal's triangle. For fixed integers a and b, define $\binom{n}{k}^b_a$ by

$$\binom{1}{0}^b_a = 1, \quad \binom{1}{1}^b_a = b, \quad \text{and} \quad \binom{1}{k}^b_a = 0 \text{ for } k \neq 0, 1;$$

$$\binom{n+1}{k}^b_a = a\binom{n}{k-1}^b_a + \binom{n}{k}^b_a \quad \text{for } n \geq 1.$$

(Note that $\binom{0}{k}^b_a$ is not defined.) Find and prove a formula for $\binom{n}{k}^b_a$. ◇

2.4 Fibonacci Numbers and Similar Recursions

We begin this section by considering the Fibonacci numbers, and a generalization of them. At the end, we will briefly consider two-term recursions more generally.

Definition 2.4.1. The *Fibonacci numbers* $F_0, F_1, F_2, \ldots$ are the sequence defined by:

$$F_0 = 0, \quad F_1 = 1, \quad \text{and } F_{n+2} = F_{n+1} + F_n \text{ for } n \geq 0.$$

Thus the Fibonacci sequence begins:

$$F_0 = 0, \quad F_1 = 1, \quad F_2 = 1, \quad F_3 = 2, \quad F_4 = 3, \quad F_5 = 5, \quad F_6 = 8,$$
$$F_7 = 13, \quad F_8 = 21, \quad F_9 = 34, \ldots.$$

We generalize this sequence as follows.

Definition 2.4.2. For arbitrary but fixed values of a and b, the sequence $G_0, G_1, G_2, \ldots$ is the sequence defined by:

$$G_0(a, b) = a, \quad G_1(a, b) = b, \quad \text{and } G_{n+2}(a, b) = G_{n+1}(a, b) + G_n(a, b) \text{ for } n \geq 0.$$

We will often abbreviate $G_n(a, b)$ to G_n.

For $a = 2$ and $b = 1$, these numbers are known as the *Lucas numbers*, and the Lucas sequence begins:

$$L_0 = 2, \ L_1 = 1, \ L_2 = 3, \ L_3 = 4, \ L_4 = 7, \ L_5 = 11, \ L_6 = 18,$$
$$L_7 = 29, \ L_8 = 47, \ L_9 = 76, \ \ldots.$$

Problem 2.4.3. Show that

$$L_n = 2F_{n-1} + F_n \quad \text{and} \quad F_n = (2/5)L_{n-1} + (1/5)L_n$$

for every $n \geq 0$. ◇

Problem 2.4.4. Define sequences $\{p_0, p_1, p_2, \ldots\}$ and $\{q_0, q_1, q_2, \ldots\}$ by

$$p_0 = 2, \qquad p_n = (1/2)p_{n-1} + (5/2)q_{n-1} \quad \text{for } n \geq 1,$$
$$q_0 = 0, \qquad q_n = (1/2)p_{n-1} + (1/2)q_{n-1} \quad \text{for } n \geq 1.$$

Show that

$$p_n = L_n \quad \text{and} \quad q_n = F_n$$

for every $n \geq 0$. ◇

Problem 2.4.5. Show that, for every positive integer n,

$$G_n(a, b) = aF_{n-1} + bF_n.$$ ◇

Problem 2.4.6. Show that, for every nonnegative integer n,

$$\sum_{k=0}^{n} G_k = G_{n+2} - G_1.$$ ◇

Problem 2.4.7. Show that, for every positive integer n,

$$\sum_{k=1}^{n} G_{2k} = G_{2n+1} - G_1.$$ ◇

Problem 2.4.8. Show that, for every positive integer n,

$$\sum_{k=1}^{n} G_{2k-1} = G_{2n} - G_0.$$ ◇

Problem 2.4.9. Show that, for every nonnegative integer n,

$$\sum_{k=0}^{n} G_k^2 = G_n G_{n+1} + G_0(G_0 - G_1).$$ ◇

Problem 2.4.10. Show that, for every $n \geq 0$,

$$G_{n+1}^2 - G_n G_{n+2} = (-1)^n (G_1^2 - G_0(G_0 + G_1)).$$ ◇

Problem 2.4.11. Show that, for every $n \geq 1$ and $m \geq 0$,

$$G_{n+m} = G_n F_{m-1} + G_{n+1} F_m.$$ ◇

As a special case of this, we have:

Corollary 2.4.12. *For every $n \geq 1$,*

$$F_{n+m} = F_n F_{m-1} + F_{n+1} F_m.$$

Proof. Since this is true for any a and b, we may take $a = 0$ and $b = 1$. □

As a further special case, we have:

Corollary 2.4.13. *For every $n \geq 1$:*

$$G_{2n-1} = G_{n-1} F_{n-1} + G_n F_n, \quad G_{2n} = G_n F_{n-1} + G_{n+1} F_n;$$
$$F_{2n-1} = F_{n-1}^2 + F_n^2, \qquad\qquad F_{2n} = (F_{n-1} + F_{n+1}) F_n.$$

Proof. Take $m = n - 1$ and $m = n$ in the above formulas. □

Problem 2.4.14. Show that, for every $n \geq 0$,

$$F_n = \sum_{k \geq 1} \binom{n-k-1}{k}.$$

(Note that this is always a finite sum.) ◇

Problem 2.4.15. Show that, for every $n \geq 0$ and $m \geq 0$,

$$G_{n+m} = \sum_{k=0}^{m} \binom{m}{k} G_{n-k}.$$ ◇

Now we turn to questions about divisibility of Fibonacci numbers.

Problem 2.4.16. (a) Show that $F_n \equiv 0 \pmod{2}$ if and only if $n \equiv 0 \pmod{3}$ and that $L_n \equiv 0 \pmod{2}$ if and only if $n \equiv 0 \pmod{3}$.

(b) Show that $F_n \equiv 0 \pmod{3}$ if and only if $n \equiv 0 \pmod{4}$ and that $L_n \equiv 0 \pmod{3}$ if and only if $n \equiv 2 \pmod{3}$. ◇

Problem 2.4.17. Suppose that a and b are relatively prime. Show that G_n and G_{n+1} are relatively prime for every $n \geq 0$. ◇

Problem 2.4.18. If k divides n, show that F_k divides F_n. ◇

Here is a more subtle result along these lines.

Problem 2.4.19. Let $d = \gcd(a, b)$. Show that $F_d = \gcd(F_a, F_b)$. ◇

Combining the last two problems, we have a stronger result.

Corollary 2.4.20. F_k divides F_n if and only if k divides n.

Proof. We have already shown that if k divides n, then F_k divides F_n. If k does not divide n, let $n = kq + r$ with $0 < r < k$. Then $F_r = \gcd(F_k, F_n)$. In particular, $\gcd(F_k, F_n) \neq F_k$, so F_k does not divide F_n. □

Problem 2.4.21. Show that, for every positive integer m, there is a positive integer n such that if $j \equiv k \pmod{n}$, then $G_j \equiv G_k \pmod{m}$. ◇

For example, in the case of the Fibonacci numbers, if $m = 2$, $n = 3$. But notice we cannot have if and only if in the above statement, as if $k \equiv 1$ or $2 \pmod 3$, then $F_k \equiv 1 \pmod 2$.

Problem 2.4.22. Show that, for every positive integer m, there is a positive integer q such that $F_k \equiv 0 \pmod{m}$ if and only if $k \equiv 0 \pmod{q}$. ◇

Note in this problem we have restricted ourselves to the Fibonacci numbers, as the result is not true for general $G_n(a, b)$.

Problem 2.4.23. Show that, for every positive integer m, there is a positive integer r such that, for any integer c, if $G_k \equiv c \pmod{m}$ and $j \equiv k \pmod{r}$, then $G_j \equiv c \pmod{m}$. ◇

Note the subtle difference between the last two problems, even in the case of the Fibonacci numbers, where they both apply. The next to the last problem is the case $c = 0$ of the last problem. But in the last problem, we allow c to be arbitrary. For example, in the case of the Fibonacci numbers, if $m = 3$ then $q = 4$, but $r = 8$.

Problem 2.4.24. Show that, for any n, the product $F_k F_{k+1} \cdots F_{k+n-1}$ of any n consecutive Fibonacci numbers is divisible by $F_1 F_2 \cdots F_n$. ◇

Note that we may regard the recursion that defines G_{n+2} in terms of G_n and G_{n+1} as defining G_n in terms of G_{n+1} and G_{n+2}, so, given G_0 and G_1, we may use this recursion to define G_n for all integers n.

Problem 2.4.25. Show that

$$G_{-n}(a, b) = (-1)^n G_n(a, a - b) \text{ for every } n \geq 0.$$ ◇

Corollary 2.4.26.

$$F_{-n} = (-1)^{n+1} F_n \text{ for every } n \geq 0.$$

Proof. Take $a = 0$ and $b = 1$, and observe that $G_n(0, -1) = -G_n(0, 1) = -F_n$ for every $n \geq 0$. □

Problem 2.4.27. Show that we have the following formula for the Fibonacci numbers:

$$F_n = \frac{1}{\sqrt{5}}\left[\left(\frac{1+\sqrt{5}}{2}\right)^n - \left(\frac{1-\sqrt{5}}{2}\right)^n\right] \text{ for every } n \geq 0. \qquad \diamond$$

This problem asks you verify the given formula. But of course you might well wonder where this formula came from, especially since it looks so odd. The Fibonacci numbers are all integers, so why should there be a formula for them that involves $\sqrt{5}$? To answer this question, we put it in a more general context.

We consider a general linear two-term recursion.

Definition 2.4.28. For fixed a, b, p, and q we define the sequence $H_0(a, b; p, q)$, $H_1(a, b; p, q)$, $H_2(a, b; p, q)$, ... by:

$$H_0(a, b; p, q) = a, \quad H_1(a, b; p, q) = b,$$
$$H_{n+2}(a, b; p, q) = pH_{n+1}(a, b; p, q) + qH_n(a, b; p, q) \text{ for } n \geq 0.$$

Of course, $H_n(a, b; 1, 1) = G_n(a, b)$, and in particular $H_n(0, 1; 1, 1) = F_n$. Let us begin by examining what happens for some small values of p and q.

Problem 2.4.29. We abbreviate $H_n(a, b; p, q)$ to $H_n(a, b)$.

(a) Let $p = -1$ and $q = -1$, so that $H_n(a, b)$ is given by $H_0(a, b) = a$, $H_1(a, b) = b$, $H_{n+2}(a, b) = -H_{n+1}(a, b) - H_n(a, b)$ for $n \geq 0$. Show that $H_{n+3}(a, b) = H_n(a, b)$ for every $n \geq 0$. Thus, in this case, the sequence $H_0(a, b)$, $H_1(a, b), H_2(a, b), \ldots$ cycles with a period of 3, and is given by $a, b, -b - a, \ldots$.

(b) Let $p = 1$ and $q = -1$, so that $H_n(a, b)$ is given by $H_0 = a$, $H_1(a, b) = b$, $H_{n+2}(a, b) = H_{n+1}(a, b) - H_n(a, b)$ for $n \geq 0$. Show that $H_{n+6}(a, b) = H_n(a, b)$ for every $n \geq 0$. Thus, in this case, the sequence $H_0(a, b)$, $H_1(a, b), H_2(a, b), \ldots$ cycles with a period of 6, and is given by $a, b, b - a, -a$, $-b, -b + a, \ldots$. $\qquad \diamond$

Problem 2.4.30. We again abbreviate $H_n(a, b; p, q)$ to $H_n(a, b)$.

(a) Let $p = -1$ and $q = 1$, so that $H_n(a, b)$ is given by $H_0(a, b) = a$, $H_1(a, b) = b$, $H_{n+2}(a, b) = -H_{n+1}(a, b) + H_n(a, b)$ for $n \geq 0$. Show that $H_n(a, b) = (-1)^n G_n(a, -b)$ for every $n \geq 0$.

(b) Show that $H_n(0, 1) = (-1)^{n+1} F_n$ for every $n \geq 0$. $\qquad \diamond$

Now we consider the general case.

Problem 2.4.31. Prove the following theorem.

Theorem 2.4.32. *Let r_1 and r_2 be the two roots of the equation $x^2 = px + q$.*

(a) If $r_1 \neq r_2$, there are unique values of x_1 and x_2 such that

$$H_n(a, b; p, q) = x_1 r_1^n + x_2 r_2^n \text{ for every } n \geq 0.$$

(b) If $r_1 = r_2 = r$, there are unique values of x_1 and x_2 such that

$$H_n(a, b; p, q) = x_1 r^n + x_2 n r^n \text{ for every } n \geq 0.$$

(Of course, in the conclusion of this theorem, x_1 and x_2 will depend not only on p and q, but also on a and b.) ◇

(There are similar theorems for linear k-term recursions for any k.)

If you carry out the proof of this theorem and solve for r_1, r_2, x_1, x_2 in the case of the Fibonacci numbers, you will arrive at the formula we gave above.

2.5 Continued Fractions

In this section, we introduce continued fraction expansions of real numbers. There is a vast theory of these, but we only go far enough to prove the basic results, with these proofs being by induction or the pigeonhole principle (of course).

Definition 2.5.1. A *continued fraction* is an expression of the form

$$a_0 + \cfrac{1}{a_1 + \cfrac{1}{a_2 + \cfrac{1}{a_3 + \cdots}}},$$

with a_0 an integer and $a_1, a_2, a_3, \ldots$ positive integers. We denote such a continued fraction by $[a_0, a_1, a_2, a_3, \ldots]$. If there are only finitely many a_i's, so that the expression is of the form $[a_0, a_1, a_2, a_3, \ldots, a_n]$, then this expression is *terminating*, and its *length* is n, while if there are infinitely many a_i's, this expression is *nonterminating*.

In developing results for continued fractions, we will often have to consider terminating expressions $[a_0, a_1, a_2, a_3, \ldots, a_{n-1}, x]$ where x is not necessarily an integer. To avoid introducing new language, we will call such an expression a continued fraction as well. (The language will be clear from the context.)

From the definition we immediately see that

$$[a_0, a_1, a_2, a_3, \ldots, a_{n-1}, y] = [a_0, a_1, a_2, a_3, \ldots, a_{n-1} + 1/y]$$

for any $y \neq 0$. This easy observation turns out to be quite useful.

Definition 2.5.2. Let x be a real number. The continued fraction expansion for x is the expansion $[a_0, a_1, \ldots]$ obtained by the following algorithm:

(i) Set $i = 0$ and $x_0 = x$.

(ii) Let a_i be the unique integer with $a_i \leq x_n < a_i + 1$. Let $r_{i+1} = x_i - a_i$. If $r_{i+1} = 0$, then $x = [a_0, \ldots, a_i]$ and we are finished. Otherwise, set $x_{i+1} = 1/r_{i+1} = 1/(x_i - a_i)$, replace i by $i + 1$, and loop.

Remark 2.5.3.

(i) Note that a_0 may be an arbitrary integer. However, for $i \geq 0$, if the algorithm has not terminated at step i, then $r_{i+1} < 1$, so $x_{i+1} = 1/r_{i+1} > 1$, and so $a_{i+1} \geq 1$. Thus we see that a_i is a positive integer for each $i \geq 1$.

(ii) Evidently, this algorithm may or may not terminate. We shall see below when it does.

(iii) Clearly the value of a terminating continued fraction expansion for x is just x. At the moment, a nonterminating continued fraction for x is just a formal expression. But we shall see (with some work) that it always converges to x.

It is worthwhile to be explicit about the intermediate steps produced by this algorithm.

Problem 2.5.4. Show that, in the notation of the above algorithm,

$$x = [a_0, a_1, \ldots, a_n, x_{n+1}]$$

if the algorithm has not terminated by step n.   ◇

Corollary 2.5.5. *Let* $x = [a_0, \ldots, a_n, x_{n+1}]$ *as above. If* x_{n+1} *has the (terminating or nonterminating) continued fraction expansion* $x_{n+1} = [b_0, b_1, \ldots]$, *then* x *has the continued fraction expansion* $[a_0, \ldots, a_n, b_0, b_1, \ldots]$.

Proof. We obtain the continued fraction expansion for x by using the algorithm up through step n, which yields $a_0, \ldots, a_n$, and then by continuing from step $n + 1$ onward, which yields $b_0, b_1, \ldots$.   □

Problem 2.5.6. (a) Let y be any positive real number. Show that, for any continued fraction expansion $[a_0, a_1, \ldots, a_n]$,

$$[a_0, a_1, \ldots, a_n] < [a_0, a_1, \ldots, a_n, y] \quad \text{for} \quad n \quad \text{even;}$$
$$[a_0, a_1, \ldots, a_n] > [a_0, a_1, \ldots, a_n, y] \quad \text{for} \quad n \quad \text{odd.}$$

(b) Let y and z be any positive real numbers with $y < z$. Show that, for any continued fraction expansion $[a_0, a_1, \ldots, a_n]$,

$$[a_0, a_1, \ldots, a_n, z] < [a_0, a_1, \ldots, a_n, y] \quad \text{for} \quad n \quad \text{even;}$$
$$[a_0, a_1, \ldots, a_n, z] > [a_0, a_1, \ldots, a_n, y] \quad \text{for} \quad n \quad \text{odd.} \quad \diamond$$

Definition 2.5.7. A *quadratic irrationality in* D is a number of the form $s + t\sqrt{D}$, where s and t are rational numbers and D is a positive integer that is not a perfect square.

As we shall soon see, there is a close connection between continued fractions and quadratic irrationalities.

It is convenient to introduce the following language: Given r, we call an expression of the form $(ar + b)/(cr + d)$, with a, b, c, d integers, a *fraction in* r.

If a, b, c, d are all nonnegative, we call this a *nonnegative fraction in r*. Note that $r = (1r + 0)/(0r + 1)$ is a nonnegative fraction in r.

Problem 2.5.8. Suppose that x is a nonnegative fraction in r. Show that the value of the expression $[a_0, a_1, a_2, a_3, \ldots, a_{n-1}, x]$ is a nonnegative fraction in r. ◇

Corollary 2.5.9. *(a) If r is a rational number, and in particular if r is an integer, then $[a_0, a_1, a_2, a_3, \ldots, a_{n-1}, r]$ is a rational number.*
(b) If r is a quadratic irrationality in D, then $[a_0, a_1, a_2, a_3, \ldots, a_{n-1}, r]$ is a quadratic irrationality in D.

Proof. We apply the previous result.

(a) If r is a rational number, then $[a_0, a_1, a_2, a_3, \ldots, a_{n-1}, r]$ is a fraction in r, and so is a rational number as well.

(b) If r is a quadratic irrationality in D, then $[a_0, a_1, a_2, a_3, \ldots, a_{n-1}, r]$ is a fraction in r, i.e., an expression of the form $(ar + b)/(cr + d)$. But then

$$\frac{ar + b}{cr + d} = \frac{a(s + t\sqrt{D}) + b}{c(s + t\sqrt{D} + d}$$
$$= \frac{a(s + t\sqrt{D}) + b}{c(s + t\sqrt{D}) + d} \cdot \frac{c(s - t\sqrt{D}) + d}{c(s - t\sqrt{D}) + d}$$
$$= \frac{((as + b) + at\sqrt{D})((cs + d) - ct\sqrt{D})}{(cs + d)^2 - t^2 D}$$

and elementary algebra shows this is a quadratic irrationality in D as well. □

Problem 2.5.10. Suppose that x is a rational number. Show that x has a terminating continued fraction expansion. ◇

We now turn our attention to nonterminating continued fraction expansions.

Definition 2.5.11. Consider a nonterminating expansion $[a_0, a_1, a_2, a_3, \ldots]$. Its nth convergent is $C_n = [a_0, a_1, a_2, a_3, \ldots, a_n]$.

Since C_n is given by a terminating continued fraction expansion, we know it is a rational number.

Problem 2.5.12. Consider a nonterminating continued fraction expansion $[a_0, a_1, a_2, a_3, \ldots]$. Let p_n and q_n be integers given by the recursion

$$p_{-2} = 0, \ p_{-1} = 1, \ p_n = a_n p_{n-1} + p_{n-2} \text{ for } n \geq 0,$$
$$q_{-2} = 1, \ q_{-1} = 0, \ p_n = a_n q_{n-1} + q_{n-2} \text{ for } n \geq 0.$$

Let C_n be the nth convergent of $[a_0, a_1, a_2, a_3, \ldots]$. Show that $C_n = p_n/q_n$ for every $n \geq 0$. ◇

Problem 2.5.13. Let p_n and q_n be as above. Show that

$$x = \frac{x_n p_{n-1} + p_{n-2}}{x_n q_{n-1} + q_{n-2}} \text{ for every } n \geq 0. \qquad \diamond$$

Problem 2.5.14. Let p_n and q_n be as above. (a) Show that

$$p_n q_{n-1} - q_n p_{n-1} = (-1)^{n-1} \text{ for every } n \geq -1.$$

(b) Show that

$$p_n q_{n-2} - q_n p_{n-2} = (-1)^n a_n \text{ for every } n \geq 0. \qquad \diamond$$

We note the following consequence of this result.

Corollary 2.5.15. *Let p_n and q_n be as above. Then p_n and q_n are relatively prime for every $n \geq -1$.*

Proof. Any common divisor of p_n and q_n would divide $p_n q_{n-1} - q_n p_{n-1} = (-1)^{n-1}$, so the only positive common divisor of p_n and q_n is 1. $\qquad \square$

Lemma 2.5.16. *Let $[a_0, a_1, \ldots]$ be any nonterminating continued fraction. Let $C_0, C_1, C_2, \ldots$ be its sequence of convergents. Then $\lim_{n \to \infty} C_n$ exists.*

Proof. Observe that, for any $n \geq 2$,

$$C_n - C_{n-2} = p_n/q_n - p_{n-2}/q_{n-2} = (p_n q_{n-2} - q_n p_{n-2})/(q_n q_{n-2})$$
$$= (-1)^n a_n/(q_n q_{n-2})$$

is positive for n even and negative for n odd. Thus the sequence of even convergents $C_0, C_2, \ldots$ is an increasing sequence while the sequence of odd convergents $C_1, C_3, \ldots$ is a decreasing sequence.

Also observe that, for any $n \geq 2$,

$$C_n - C_{n-1} = p_n/q_n - p_{n-1}/q_{n-1} = (p_n q_{n-1} - q_n p_{n-1})/(q_n q_{n-1})$$
$$= (-1)^{n-1}/(q_n q_{n-1}).$$

Recall the recursion for q_n: $q_0 = 1$, $q_1 = 1$, and $q_{n+1} = a_n q_n = q_{n-1}$. Since $a_n \geq 1$ for $n \geq 1$, we immediately see that $q_n \geq n$ for every $n \geq 1$. Thus we see that

$$|C_n - C_{n-1}| \leq 1/(n(n-1)) \text{ for every } n \geq 2.$$

These two observations imply that the sequences $C_0, C_2, \ldots$ and $C_1, C_3, \ldots$ have a common limit, i.e., that the sequence $C_0, C_1, C_2, \ldots$ has a limit. $\qquad \square$

Given this lemma, we are justified in making the following definition.

Definition 2.5.17. The value x of a continued fraction $[a_0, a_1, \ldots]$ is

$$x = \lim_{n \to \infty} C_n = \lim_{n \to \infty} [a_0, a_1, \ldots, a_n].$$

Now we have developed the continued fraction expansion for any real number x. We have also shown that every continued fraction expansion has some value. There is one thing left to show: that the value of the continued fraction expansion for x is indeed x. We show that now.

Theorem 2.5.18. *Let x have the continued fraction expansion $[a_0, a_1, \ldots]$. Then the value of this continued fraction expansion is x.*

First Proof. This is trivially true when the expansion terminates as, if it has length n, then $C_n = x$. So we need only show this for nonterminating expansions.

Recall that we have

$$x = [a_0, \ldots, a_n, x_{n+1}]$$

for any n, so by the above result

$$x > [a_0, \ldots, a_n] = C_n \text{ for } n \text{ even}, \quad x < [a_0, \ldots, a_n] = C_n \text{ for } n \text{ odd}.$$

Thus we see that

$$\lim_{n \text{ even}} C_n \le x \text{ and } \lim_{n \text{ odd}} C_n \ge x.$$

But we have already seen that these two limits are the same. Hence their common value must be x. $\qquad\qquad\square$

Second Proof. We have that

$$
\begin{aligned}
x - C_n &= \frac{x_{n+1} p_n + p_{n-1}}{x_{n+1} q_n + q_{n-1}} - \frac{p_n}{q_n} \\
&= \frac{q_n (x_{n+1} p_n + p_{n-1}) - p_n (x_{n+1} q_n + q_{n-1})}{q_n (x_{n+1} q_n + q_{n-1})} \\
&= \frac{q_n p_{n-1} - p_n q_{n-1}}{q_n (x_{n+1} q_n + q_{n-1})} \\
&= \frac{(-1)^n}{q_n (x_{n+1} q_n + q_{n-1})}.
\end{aligned}
$$

But, as in the first proof, $q_n \ge n$ for every $n \ge 1$, and also $x_n \ge 1$ for every $n \ge 1$, so we see

$$|x - C_n| \le \frac{1}{n(2n - 1)}$$

for every $n \ge 1$, and that immediately implies $\lim_{n \to \infty} C_n = x$. $\qquad\qquad\square$

Definition 2.5.19. A continued fraction expansion $[a_0, a_1, a_2, a_3, \ldots]$ is *periodic* of period m beginning with a_k if $a_{j+m} = a_j$ for every $j \ge k$, i.e., if

$$[a_0, a_1, a_2, a_3, \ldots]$$
$$= [a_0, a_1, a_2, a_3, \ldots, a_{k-1}, a_k, \ldots, a_{k+m-1}, a_k, \ldots, a_{k+m-1}, \ldots].$$

If $k = 0$, so that the periodicity begins with a_0, the expansion is *purely periodic*.

Theorem 2.5.20. *Let x have a periodic continued fraction expansion. Then x is a quadratic irrationality.*

Proof. Let $x = [a_0, a_1, a_2, a_3, \ldots]$. First suppose this expansion is purely periodic. Then for some m,

$$x = [a_0, \ldots, a_{m-1}, a_0, \ldots, a_{m-1}, a_0, \ldots, a_{m-1}, \ldots]$$
$$= [a_0, \ldots, a_{m-1}, x].$$

But by a previous result, $[a_0, \ldots, a_{m-1}, x]$ is a nonnegative fraction in x, i.e., is equal to $(ax + b)/(cx + d)$ for some a, b, c, d all nonnegative integers. Thus we see

$$x = (ax + b)/(cx + d) \quad \text{so} \quad cx^2 + dx = ax + b$$
and so $\quad cx^2 + (d - a)x - b = 0.$

But this is a quadratic equation, and the nonnegativity of a, b, c, d implies that it has two real roots, one positive and one negative. If x is positive, it must be equal to the positive root, and if x is negative, it must be equal to the negative root. But from the quadratic formula, this equations has roots

$$x = s + t\sqrt{D} \quad \text{where} \quad s = (a - d)/(2c), \ t = \pm 1/(2c),$$
and $\quad D = (d - a)^2 + 4bc$

both of which are evidently quadratic irrationalities.

Now suppose that the expansion is not purely periodic and instead that the period begins at a_k. Let $y = [a_k, \ldots, a_{k+m-1}, a_k, \ldots, a_{k+m-1}, \ldots]$. This is a purely periodic expansion and so y is a quadratic irrationality. But then $x = [a_0, \ldots, a_{k-1}, y]$ is a fraction in y and so is a quadratic irrationality as well. $\square$

We now do some illustrative computations of continued fractions. Recall that, in our above notation, when computing the continued fraction expansion of the real number x, we have that

$$x = [x_0] = [a_0, x_1] = [a_0, a_1, x_2] = [a_0, a_1, a_2, x_3] = \ldots$$

Example 2.5.21. (a) First we compute the continued fraction expansion of the rational number $x = 100/27$. Of course, we know this expansion will terminate.

$$x_0 = 100/27 \text{ so } a_0 = 3, \quad r_1 = 19/27, \quad x_1 = 27/19.$$
Thus $100/27 = [3, 27/19]$.

$$x_1 = 27/19 \text{ so } a_1 = 1, \quad r_2 = 8/19, \quad x_2 = 19/8.$$
Thus $100/27 = [3, 1, 19/8]$.

$$x_2 = 19/8 \text{ so } a_2 = 2, \quad r_3 = 3/8, \quad x_3 = 8/3.$$

Thus $100/27 = [3, 1, 2, 8/3]$.

$$x_3 = 8/3 \text{ so } a_3 = 2, \quad r_4 = 2/3, \quad x_4 = 3/2.$$

Thus $100/27 = [3, 1, 2, 2, 3/2]$.

$$x_4 = 3/2 \text{ so } a_4 = 1, \quad r_5 = 1/2, \quad x_5 = 2.$$

Thus $100/27 = [3, 1, 2, 2, 1, 2]$.

Now x_5 is an integer, so we are done, and so this is the continued fraction expansion of $100/27$.

(b) Next we compute the continued fraction expansion of $(13 - \sqrt{17})/4$. Since this number is irrational, we know that its continued fraction expansion will not terminate.

$$x_0 = (13 - \sqrt{17})/4 \quad \text{so} \quad a_0 = 2, \quad r_1 = (5 - \sqrt{17})/4, \quad x_1 = 4/(5 - \sqrt{17}).$$

But then

$$x_1 = \frac{4}{5 - \sqrt{17}} = \frac{4}{5 - \sqrt{17}} \cdot \frac{5 + \sqrt{17}}{5 + \sqrt{17}} = \frac{5 + \sqrt{17}}{2}.$$

Thus $(13 - \sqrt{17})/4 = [2, (5 + \sqrt{17})/2]$.

$x_1 = (5 + \sqrt{17})/2$ so $a_1 = 4, \quad r_2 = (-3 + \sqrt{17})/2, \quad x_2 = 2/(-3 + \sqrt{17})$.

But then

$$x_2 = \frac{2}{-3 + \sqrt{17}} = \frac{2}{-3 + \sqrt{17}} \cdot \frac{-3 - \sqrt{17}}{-3 - \sqrt{17}} = \frac{3 + \sqrt{17}}{4}.$$

Thus $(13 - \sqrt{17})/4 = [2, 4, (3 + \sqrt{17})/4]$.

$x_2 = (3 + \sqrt{17})/4$ so $a_2 = 1, \quad r_3 = (-1 + \sqrt{17})/4, \quad x_3 = 4/(-1 + \sqrt{17})$.

But then

$$x_3 = \frac{4}{-1 + \sqrt{17}} = \frac{4}{-1 + \sqrt{17}} \cdot \frac{-1 - \sqrt{17}}{-1 - \sqrt{17}} = \frac{1 + \sqrt{17}}{4}.$$

Thus $(13 - \sqrt{17})/4 = [2, 4, 1, (1 + \sqrt{17})/4]$.

$x_3 = (1 + \sqrt{17})/4$ so $a_3 = 1, \quad r_4 = (-3 + \sqrt{17})/4, \quad x_4 = 4/(-3 + \sqrt{17})$.

But then

$$x_4 = \frac{4}{-3 + \sqrt{17}} = \frac{4}{-3 + \sqrt{17}} \cdot \frac{-3 - \sqrt{17}}{-3 - \sqrt{17}} = \frac{3 + \sqrt{17}}{2}.$$

Thus $(13 - \sqrt{17})/4 = [2, 4, 1, 1, (3 + \sqrt{17})/2]$.

$x_4 = (3 + \sqrt{17})/2$ so $a_4 = 3$, $r_5 = (-3 + \sqrt{17})/2$, $x_5 = 2/(-3 + \sqrt{17})$.

But then

$$x_5 = \frac{2}{-3 + \sqrt{17}} = \frac{2}{-3 + \sqrt{17}} \cdot \frac{-3 - \sqrt{17}}{-3 - \sqrt{17}} = \frac{3 + \sqrt{17}}{4}.$$

Thus $(13 - \sqrt{17})/4 = [2, 4, 1, 1, 3, (3 + \sqrt{17})/4]$.

We observe that $x_5 = x_2$. But then $a_5 = a_2$, $r_6 = r_3$, $x_6 = x_3$, etc., and the sequence cycles forever.

Thus $(13 - \sqrt{17})/4 = [2, 4, 1, 1, 3, 1, 1, 3, 1, 1, 3, \ldots]$.

We see that this continued fraction expansion is periodic of period 3 beginning with a_2.

We write $(13 - \sqrt{17})/4 = [2, 4, \overline{1, 1, 3}]$ where the overbar denotes the periodic part. (This is standard notation.)

(c) We compute the continued fraction expansion of $\sqrt{13}$. Here (and henceforth) we will not give the intermediate computations for x_n but just give the final answer.

$x_0 = \sqrt{13}$ so $a_0 = 3$, $r_1 = -3 + \sqrt{13}$, $x_1 = (3 + \sqrt{13})/4$.

Thus $\sqrt{13} = [3, (3 + \sqrt{13})/4]$.

$x_1 = (3 + \sqrt{13})/4$ so $a_1 = 1$, $r_2 = (-1 + \sqrt{13})/4$, $x_2 = (1 + \sqrt{13})/3$.

Thus $\sqrt{13} = [3, 1, (1 + \sqrt{13})/3]$.

$x_2 = (1 + \sqrt{13})/3$ so $a_2 = 1$, $r_3 = (-2 + \sqrt{13})/4$, $x_3 = (2 + \sqrt{13})/3$.

Thus $\sqrt{13} = [3, 1, 1, (2 + \sqrt{13})/3]$.

$x_3 = (2 + \sqrt{13})/3$ so $a_3 = 1$, $r_4 = (-1 + \sqrt{13})/3$, $x_4 = (1 + \sqrt{13})/4$.

Thus $\sqrt{13} = [3, 1, 1, 1, (1 + \sqrt{13})/4]$.

$x_4 = (1 + \sqrt{13})/4$ so $a_4 = 1$, $r_5 = (-3 + \sqrt{13})/4$, $x_5 = 3 + \sqrt{13}$.

Thus $\sqrt{13} = [3, 1, 1, 1, 1, 3 + \sqrt{13}]$.

$x_5 = 3 + \sqrt{13}$ so $a_5 = 6$, $r_6 = -3 + \sqrt{13}$, $x_6 = (3 + \sqrt{13})/4$.

Thus $\sqrt{13} = [3, 1, 1, 1, 6, (3 + \sqrt{13})/4]$.

We observe that $x_6 = x_1$ so at this point the sequence cycles.

Thus $\sqrt{13} = [3, \overline{1, 1, 1, 1, 6}]$.

We see that this continued fraction expansion is periodic of period 5 beginning with a_1.

(d) We compute the continued fraction expansion of $\sqrt{19}$.

$$x_0 = \sqrt{19} \quad \text{so} \quad a_0 = 4, \quad r_1 = -4 + \sqrt{19}, \quad x_1 = (4 + \sqrt{19})/3.$$

Thus $\sqrt{19} = [4, (4 + \sqrt{19})/3]$.

$$x_1 = (4 + \sqrt{19})/3 \quad \text{so} \quad a_1 = 2, \quad r_2 = (-2 + \sqrt{19})/5, \quad x_2 = (2 + \sqrt{19})/5.$$

Thus $\sqrt{19} = [4, 2, (2 + \sqrt{19})/5]$.

$$x_2 = (2 + \sqrt{19})/5 \quad \text{so} \quad a_2 = 1, \quad r_3 = (-3 + \sqrt{19})/5, \quad x_3 = (3 + \sqrt{19})/2.$$

Thus $\sqrt{19} = [4, 2, 1, (3 + \sqrt{19})/2]$.

$$x_3 = (3 + \sqrt{19})/2 \quad \text{so} \quad a_3 = 3, \quad r_4 = (-2 + \sqrt{19})/5, \quad x_4 = (3 + \sqrt{19})/5.$$

Thus $\sqrt{19} = [4, 2, 1, 3, (3 + \sqrt{19})/5]$.

$$x_4 = (3 + \sqrt{19})/5 \quad \text{so} \quad a_4 = 1, \quad r_5 = (-2 + \sqrt{19})/5, \quad x_5 = (2 + \sqrt{19})/3.$$

Thus $\sqrt{19} = [4, 2, 1, 3, 1, (2 + \sqrt{19})/3]$.

$$x_5 = (2 + \sqrt{19})/3 \quad \text{so} \quad a_5 = 2, \quad r_6 = (-4 + \sqrt{19})/3, \quad x_6 = 4 + \sqrt{19}.$$

Thus $\sqrt{19} = [4, 2, 1, 3, 1, 2, 4 + \sqrt{19}]$.

$$x_6 = 4 + \sqrt{19} \quad \text{so} \quad a_6 = 8, \quad r_7 = -4 + \sqrt{19}, \quad x_7 = (4 + \sqrt{19})/3.$$

Thus $\sqrt{19} = [4, 2, 1, 3, 1, 2, 8, (4 + \sqrt{19})/3]$.
We observe that $x_7 = x_1$ so at this point the sequence cycles.
Thus $\sqrt{19} = [4, \overline{2, 1, 3, 1, 2, 8}]$.

We see that this continued fraction expansion is periodic of period 6 beginning with a_1.

2.6 Polynomials and Other Functions

In this section we present a number of problems, mostly about polynomials, but also some about other functions as well. In this section, a function will always mean a function from the real numbers to the real numbers.

Recall that a polynomial $p(x) = a_0 + a_1 x + a_2 x^2 \ldots + a_n x^n$ with $a_n \neq 0$ has *degree* n. In particular, a polynomial of degree 0 is a *nonzero* constant polynomial. We define the degree of the 0 polynomial to be $-\infty$, and adopt the convention that $-\infty < n$ and $-\infty + n = -\infty$ for any nonnegative integer n. (With this convention, $\deg(f(x)g(x)) = \deg(f(x)) + \deg(g(x))$ for any two polynomials $f(x)$ and $g(x)$; without this convention we would have to make a special case for the 0 polynomial.)

We begin with some fundamental properties of polynomials.

Problem 2.6.1. Show that the analog of the division algorithm holds for polynomials: If $a(x)$ is a polynomial and $b(x)$ is a nonzero polynomial, then there are unique polynomials $q(x)$ and $r(x)$ with

$$a(x) = b(x)q(x) + r(x) \text{ with } r(x) = 0 \text{ or } \deg(r(x)) < \deg(b(x)). \quad \diamond$$

One very special case of this theorem is already important.

Corollary 2.6.2. *Let $p(x)$ be an arbitrary polynomial. Then for any a, $p(x) = (x-a)q(x) + p(a)$, for some unique polynomial $q(x)$. In particular, $p(x)$ is divisible by $x - a$ if and only if $p(a) = 0$, i.e., if and only if a is a root of $p(x)$.*

Proof. By the above problem, $p(x) = (x-a)q(x) + c$ for some constant c. Setting $x = a$ in this equation, we see $c = p(a)$.   □

Problem 2.6.3. Show that a polynomial of degree d has at most d roots. $\diamond$

Problem 2.6.4. Let $p(x_1, \ldots, x_k)$ be a nonzero polynomial in k variables. Show that $p(a_1, \ldots, a_k) \neq 0$ for some $a_1, \ldots, a_k$. $\diamond$

Problem 2.6.5. Show there is a sequence of real numbers $a_0, a_1, \ldots$ such that for every $n \geq 0$, the polynomial $p_n(x) = a_0 + a_1 x + \ldots + a_n x^n$ has n distinct real roots. $\diamond$

For $n \geq 0$ we define the polynomial $x^{(n)}$ by

$$x^{(n)} = x(x-1)\ldots(x-n+1)$$

so that the first few of these are given by

$$x^{(0)} = 1, \quad x^{(1)} = x, \quad x^{(2)} = x(x-1), \quad x^{(3)} = x(x-1)(x-2),$$

and we see that $x^{(n)}$ is a polynomial of degree n.

Clearly, for each n, $x^{(n)}$ is a polynomial with integer coefficients, i.e., for each n

$$x^{(n)} = \sum_{k=0}^{n} a_{n,k} x^k \text{ for unique integers } a_{n,0}, a_{n,1}, \ldots, a_{n,n}.$$

In other words, the polynomials $x^{(n)}$ can be uniquely expressed in terms of the polynomials x^k.

It is also not hard to see that the reverse is true. That is, for each n,

$$x^n = \sum_{k=0}^{n} b_{n,k} x^{(k)} \text{ for unique integers } b_{n,0}, b_{n,1}, \ldots, b_{n,n}.$$

In other words, the polynomials x^n can be uniquely expressed in terms of the polynomials $x^{(k)}$.

It is natural to ask about the values of the integers $\{a_{n,k}\}$ and $\{b_{n,k}\}$ and we will consider these below.

Problem 2.6.6. Show that

$$\sum_{j=0}^{k} j^{(n)} = \frac{(k+1)^{(n+1)}}{n+1}$$

for all nonnegative integers k and n. $\diamond$

Corollary 2.6.7. *For every integer $n \geq 0$, there is a polynomial $P_{n+1}(x)$ of degree $n+1$ such that*

$$\sum_{j=1}^{k} j^n = P_{n+1}(k)$$

for every integer $k \geq 1$.
Furthermore,

(a) $P_{n+1}(x)$ has rational coefficients;

(b) the coefficient of x^{n+1} in $P_{n+1}(x)$ is $\frac{1}{n+1}$;

(c) $P_{n+1}(x)$ is divisible by x (so that the constant term of $P_{n+1}(x)$ is 0); and

(d) $P_{n+1}(x)$ is divisible by $x+1$ for every $n \geq 1$.

Proof. We prove this by complete induction on n. For $n = 0$ we have $P_1(x) = x$, and, as we have already seen, for $n = 1$ we have $P_2(x) = x(x+1)/2$.

For ease of notation, let us set $Q_{n+1}(x) = \frac{(x+1)^{(n+1)}}{n+1}$. Observe that $Q_{n+1}(x)$ satisfies properties (a) through (d) above.

As we have just seen, $\sum_{j=0}^{k} j^{(n)} = Q_{n+1}(k)$ for every $n \geq 0$. But $j^{(n)}$ is a polynomial of degree n in j, so we may expand it as

$$j^{(n)} = c_n j^n + c_{n-1} j^{n-1} + \ldots + c_1 j + c_0.$$

We see immediately that each c_i is an integer, and that $c_n = 1$ and $c_0 = 0$. Thus

$$Q_{n+1}(k) = \sum_{j=0}^{k} j^{(n)}$$

$$= \sum_{j=0}^{k} j^n + c_{n-1} j^{n-1} + \ldots + c_1 j$$

$$= \sum_{j=0}^{k} j^n + c_{n-1} \sum_{j=0}^{k} j^{n-1} + \ldots + c_1 \sum_{j=0}^{k} j$$

$$= \sum_{j=0}^{k} j^n + c_{n-1} P_n(k) + \ldots + c_1 P_2(k)$$

by the inductive hypothesis. But then $\sum_{j=0}^{k} j^n = Q_{n+1}(k) - c_{n-1} P_n(k) - \ldots - c_1 P_2(k)$ so

$$P_{n+1}(x) = Q_{n+1}(x) - c_{n-1} P_n(x) - \ldots - c_1 P_2(x)$$

is the required polynomial, and by induction this is the case for every $n \geq 0$. $\quad\square$

Here is a similar identity.

Problem 2.6.8. Show that

$$\sum_{j=n+1}^{n+k} \frac{1}{j^{(n+1)}} = \frac{1}{n} \left(\frac{1}{n!} - \frac{1}{(n+k)^{(n)}} \right)$$

for all positive integers k and n. $\quad\diamond$

Problem 2.6.9. Prove the following analog of the binomial theorem: For every nonnegative integer n,

$$(x + y)^{(n)} = \sum_{k=0}^{n} \binom{n}{k} x^{(n-k)} y^{(k)}.\qquad\diamond$$

Problem 2.6.10. The *Gamma function* is defined by

$$\Gamma(x) = \int_0^\infty t^{x-1} e^{-t} dt \text{ for } x > 0.$$

Show that

$$\Gamma(x + n) = x(x + 1)\ldots(x + (n - 1))\Gamma(x) = (x + (n - 1))^{(n)}\Gamma(x)$$

for every positive integer n. $\quad\diamond$

Corollary 2.6.11. $\Gamma(n) = (n - 1)!$ *for every positive integer n.*

Proof. By its definition, and elementary calculus

$$\Gamma(1) = \int_0^\infty e^{-t} dt = -e^{-t} \Big|_0^\infty = 1 = 0!$$

and then, by the above problem,

$$\Gamma(n) = \Gamma(1 + (n - 1)) = (1 + (n - 2))^{(n-1)}\Gamma(1)$$
$$= (n - 1)^{(n-1)}\Gamma(1) = (n - 1)! \cdot 1 = (n - 1)!$$

for every $n \geq 2$. $\quad\square$

Problem 2.6.12. (a) Show that

$$n! > n^n e^{-n} \text{ for every } n \geq 1.$$

(b) Show that

$$n! < n^{n+1}e^{-n} \text{ for every } n \geq 7. \qquad \diamond$$

In fact Stirling's formula shows that $n!$ is asymptotic to $n^n e^{-n}\sqrt{2\pi n}$, i.e.,
that

$$\lim_{n \to \infty} \frac{n!}{n^n e^{-n}\sqrt{2\pi n}} = 1.$$

We let I be the identity operator on functions ($I(f(x)) = f(x)$), we define
the *first forward translation operator* T on functions by $T(f(x)) = f(x+1)$, and
we define the *first forward difference operator* Δ on functions by $\Delta = T - I$, i.e.,

$$\Delta(f(x)) = (T - I)(f(x)) = f(x+1) - f(x).$$

Problem 2.6.13. Let $f(x)$ be an arbitrary function. Show that

$$\Delta^n(f(x)) = \sum_{k=0}^{n}(-1)^{n-k}\binom{n}{k}f(x+k). \qquad \diamond$$

Problem 2.6.14. Prove the following: Let $f(x)$ be an arbitrary function. Then
$f(x)$ is a polynomial of degree n if and only if $i = n+1$ is the smallest positive
integer for which $\Delta^i(f(x)) = 0$. $\qquad \diamond$

Problem 2.6.15. Prove *Newton's identity:*
Let $f(x)$ be a polynomial of degree n. Then

$$f(x) = \sum_{i=0}^{n} \frac{\Delta^i(f)(0)}{i!}x^i.$$

Here $\Delta^i(f)(0)$ means the value of the function $\Delta^i(f(x))$ at $x = 0$. $\qquad \diamond$

Problem 2.6.16. Prove *Descartes' rule of signs:*
Let $f(x)$ be a polynomial of degree n with real coefficients, $f(x) = a_n x^n + a_{n-1}x^{n-1} + \ldots + a_0$. Write the coefficients $a_n, a_{n-1}, \ldots, a_0$ in order, omitting any
coefficients that are equal to 0. Let c be the number of sign changes in this sequence,
i.e., the number of times that two consecutive coefficients are of opposite signs.
Then the number p of positive real roots of $f(x) = 0$ (counted with multiplicity)
is less than or equal to c. $\qquad \diamond$

Actually, Descartes's rule of signs says a bit more. Note that if c is even, then
the first term a_n in this sequence and the last term a_k in this sequence have the same
sign, while if c is odd, then a_n and a_k have opposite signs. (This last term is a_0
if 0 is not a root of $f(x)$, of course.) Let $f(x)$ have positive real roots $r_1, \ldots, r_p$,
and negative real roots $-s_1, \ldots, -s_q$. Then

$$f(x) = x^k(x - r_1) \cdots (x - r_p)(x + s_1) \cdots (x + s_q)g(x)$$

where $g(x)$ is a polynomial of degree $m = n - (p + q + k)$ with no real roots.
Then $g(x) = b_m x^m + \ldots + b_0$ is a polynomial, necessarily of even degree, with b_m

and b_0 having the same sign. But $a_n = b_m$ and $a_k = (-1)^p r_1 \cdots r_p s_1 \cdots s_q b_0$ so a_n and a_k have the same sign if p is even and opposite signs if p is odd. Thus we can conclude that the number p of real roots of $f(x) = 0$ is equal to $c - 2d$ for some nonnegative integer d.

Here is another analog of the binomial theorem. For a function $f(x)$, we let $D(f(x)) = f'(x)$ be its derivative and then $D^n(f(x)) = f^{(n)}(x)$ is its nth derivative.

Problem 2.6.17. Let $f(x)$ and $g(x)$ be any two functions that have derivatives of all orders. Show that, for every $n \geq 0$,

$$D^n(f(x)g(x)) = \sum_{k=0}^{n} \binom{n}{k} D^{n-k}(f(x))D^k(g(x)). \qquad \diamond$$

Problem 2.6.18. Let $f(x)$ be a function and suppose that $f^{(k)}(a) = 0$ for $k = 0, \ldots, n-1$ and that $f^{(n)}(a)$ exists. Show that

$$\lim_{x \to a} \frac{f(x)}{(x-a)^n} = \frac{f^{(n)}(a)}{n!}. \qquad \diamond$$

This result has the following consequence. Note that in this theorem, $E_n(x)$ is the error in approximating $f(x)$ by its nth order Taylor polynomial at $x = a$, so this theorem states that this error approaches 0 very quickly as x approaches a.

Theorem 2.6.19. *Let $f(x)$ be a function that is n-times differentiable at $x = a$. Let $T_n(x)$ be the nth order Taylor polynomial of $f(x)$ at $x = a$,*

$$T_n(x) = f(a) + f'(a)(x-a) + \ldots + \frac{f^{(n)}(a)}{n!}(x-a)^n$$

and let

$$E_n(x) = f(x) - T_n(x).$$

Then

$$\lim_{x \to a} \frac{E_n(x)}{(x-a)^n} = 0.$$

Proof. Let $T_{n-1}(x) = f(a) + f'(a)(x-a) + \ldots + \frac{f^{(n-1)}(a)}{(n-1)!}(x-a)^{n-1}$ be the $(n-1)$st order Taylor polynomial and let $h(x) = f(x) - T_{n-1}(x)$. Then, we have that

$$h^{(k)}(a) = 0, \quad k = 0, \ldots, n-1, \quad h^{(n)}(a) = f^{(n)}(a),$$

$$\text{and} \quad h(x) = \frac{f^{(n)}(a)}{n!}(x-a)^n + E_n(x).$$

Then, on the one hand, from the previous problem we have that

$$\lim_{x \to a} \frac{h(x)}{(x-a)^n} = \frac{h^{(n)}(a)}{n!} = \frac{f^{(n)}(a)}{n!}$$

and on the other hand, we have that

$$\lim_{x \to a} \frac{h(x)}{(x-a)^n} = \lim_{x \to a} \frac{\frac{f^{(n)}(a)}{n!}(x-a)^n + E_n(x)}{(x-a)^n}$$

$$= \lim_{x \to a} \frac{f^{(n)}(a)}{n!} + \lim_{x \to a} \frac{E_n(x)}{(x-a)^n}$$

$$= \frac{f^{(n)}(a)}{n!} + \lim_{x \to a} \frac{E_n(x)}{(x-a)^n},$$

and so comparing these expressions we see that $\lim_{x \to a} \frac{E_n(x)}{(x-a)^n} = 0$. $\square$

Problem 2.6.20. Let $f(x)$ be a function whose $(n-1)$st derivative is continuous on the closed interval $[a, b]$, and whose nth derivative exists at every point in the open interval (a, b), $n \geq 1$. Suppose that $f^{(k)}(a) = 0$ for $k = 1, \ldots, n-1$ and that $f(b) = f(a)$. Show that there is some point c in the interval (a, b) with $f^{(n)}(c) = 0$. $\diamond$

This result, which is a generalization of Rolle's theorem (the case $n = 1$), has the following consequence, which is a generalization of the mean value theorem (also the case $n = 1$).

Theorem 2.6.21. Let $f(x)$ be a function whose $(n-1)$st derivative is continuous on the closed interval $[a, b]$, and whose nth derivative exists at every point in the open interval (a, b), $n \geq 1$. Let $T_{n-1}(x)$ be the $(n-1)$st order Taylor polynomial of $f(x)$ at $x = a$. Then for some point c in the interval (a,b),

$$f(b) = T_{n-1}(b) + \frac{f^{(n)}(c)}{n!}(b-a)^n.$$

Proof. Let $F(x)$ be the function $F(x) = f(x) - T_{n-1}(x)$, and note that $F^{(k)}(a) = 0$ for $0 \leq k \leq n-1$. Let $G(x)$ be the function

$$G(x) = F(x) - \frac{F(b)}{(b-a)^n}(x-a)^n.$$

Then $G^{(k)}(a) = F^{(k)}(a) = 0$ for $0 \leq k \leq n-1$, and $G(b) = G(a) = 0$. Hence by the previous problem there is a point c in the interval (a, b) with $G^{(n)}(c) = 0$. But then

$$0 = G^{(n)}(c) = F^{(n)}(c) - \frac{F(b)}{(b-a)^n}n!$$

from which we conclude

$$F(b) = \frac{F^{(n)}(c)}{n!}(b-a)^n.$$

But $F(b) = f(b) - T_{n-1}(b)$ so

$$f(b) = T_{n-1}(b) + \frac{f^{(n)}(c)}{n!}(b - a)^n$$

as claimed.   □

This immediately yields an upper bound for the error $E_n(x)$.

Corollary 2.6.22. *Let $f(x)$ be a function satisfying the hypotheses of the previous theorem. Suppose that*

$$|f^{(n)}(c) - f^{(n)}(a)| \le M \text{ for every } c \text{ in } (a, b).$$

Then

$$|E_n(x)| = |f(x) - T_n(x)| \le (M/n!)|x - a|^n$$

for every x in the interval $[a, b]$.

Replacing $n - 1$ by n in this theorem, leaving a fixed, and letting the point at which we are evaluating the function vary, we obtain the following result, which is Taylor's theorem with the Lagrange form of the remainder.

Theorem 2.6.23. *Let $f(x)$ be a function whose $(n-1)$st derivative is continuous on the closed interval $[a, b]$, and whose nth derivative exists at every point in the open interval (a, b), $n \ge 1$. Let $T_n(x)$ be the nth order Taylor polynomial of $f(x)$ at $x = a$. Then for any point x in the interval $[a, b]$,*

$$f(x) = f(a) + f'(a)(x - a) + \ldots + \frac{f^{(n)}(a)}{n!}(x - a)^n + \frac{f^{(n+1)}(c)}{(n+1)!}(x - a)^{n+1}$$

for some point c in the interval (a,b).

Under these hypotheses we obtain a better upper bound on the error term $E_n(x)$ in the Taylor expansion. Note these hypotheses are stronger than our previous hypotheses in that we require one more derivative for the function $f(x)$.

Corollary 2.6.24. *Let $f(x)$ be a function satisfying the hypotheses of the previous theorem. Suppose that*

$$|f^{(n+1)}(c)| \le M \text{ for every } c \text{ in } (a, b).$$

Then

$$|E_n(x)| = |f(x) - T_n(x)| \le (M/(n+1)!)|x - a|^{n+1}$$

for every x in the interval $[a, b]$.

Problem 2.6.25. For $k \ge 0$ an integer, let $f_k(z)$ be the function $f_k(z) = 1/(1-z)^{k+1}$. Show that $f_k(z)$ has the power series expansion, valid for $|z| < 1$,

$$f_k(z) = \sum_{n=0}^{\infty} \binom{n+k}{k} z^n. \qquad \diamond$$

Problem 2.6.26. Let $k \geq 0$ be an integer and let $x > 1$ be a rational number. Let

$$s(k, x) = \sum_{n=0}^{\infty} \frac{n^k}{x^n}.$$

Show that $s(k, x)$ is a rational number. ◇

Let us define an *integer-valued polynomial* to be a polynomial $p(x)$ with the property that $p(n)$ is an integer for every integer n. Of course, any polynomial with integer coefficients is an integer-valued polynomial. But these are not the only ones. We observe that for any integer x, $x^{(n)}$ is the product of n consecutive integers, so by a previous result is always divisible by $n!$ (to be precise, we proved this for a product of n consecutive positive integers, but then it immediately follows for a product of any n consecutive integers); consequently $x^{(n)}/n!$ is an integer-valued polynomial.

Problem 2.6.27. Let $p(x)$ be a polynomial of degree at most n. If $p(x)$ is an integer for $n + 1$ consecutive integer values of x, prove that $p(x)$ is an integer-valued polynomial. ◇

Problem 2.6.28. Prove the following theorem of Polya:

Theorem 2.6.29. Let $p(x)$ be an integer-valued polynomial of degree n. Then

$$p(x) = \sum_{k=0}^{n} a_k \frac{x^{(k)}}{k!}$$

where $a_0, a_1, \ldots, a_n$ are all integers. ◇

Problem 2.6.30. Prove Hensel's lemma:

Theorem 2.6.31. Let p be a prime. Let $s_1, s_2, s_3, \ldots$ be a sequence of integers with $s_k \equiv s_{k-1} \pmod{p^{k-1}}$ for every $k \geq 2$. Let $f(x)$ be a polynomial with integer coefficients. Suppose there is an integer r_1 with $f(r_1) \equiv s_1 \pmod{p}$ and with $f'(r_1) \not\equiv 0 \pmod{p}$. Show that for every $k \geq 1$ there is an integer r_k with $f(r_k) \equiv s_k \pmod{p^k}$. ◇

Note that the conditions in Hensel's lemma are in general necessary. We need to assume there is an integer r_1 with $f(r_1) \equiv s_1 \pmod{p}$ to get started. For example, the congruence $x^2 \equiv 2 \pmod{3}$ has no solution. And we need to assume $f'(r_1) \not\equiv 0 \pmod{p}$ to be able to continue. For example, if $s_1 = 1$, $s_2 = 3$ the congruence $x^2 \equiv 1 \pmod{2}$ has a solution but the congruence $x^2 \equiv 3 \pmod{4}$ does not.

Observe that

$$\cos(x) = \cos(x), \qquad \sin(x) = \sin(x)[1];$$
$$\cos(2x) = 2\cos^2(x) - 1, \qquad \sin(2x) = \sin(x)[2\cos(x)];$$
$$\cos(3x) = 4\cos^3(x) - 3\cos(x)), \qquad \sin(3x) = \sin(x)[4\cos^2(x) - 1].$$

We see that, in these cases, $\cos(nx)$ is a polynomial of degree n in $\cos(x)$, and $\sin(nx)$ is the product of $\sin(x)$ and a polynomial of degree $n-1$ in $\cos(x)$. In fact, this pattern continues forever.

Problem 2.6.32. Let $T_0(x) = 1$. Let $T_1(x) = x$ and $U_0(x) = 1$, and, given $T_n(x)$ and $U_{n-1}(x)$, let

$$T_{n+1}(x) = xT_n(x) - (1 - x^2)U_{n-1}(x), \qquad U_n(x) = xU_{n-1}(x) + T_n(x),$$

and note that $T_n(x)$ is a polynomial of degree n and also that $U_{n-1}(x)$ is a polynomial of degree $n - 1$.
    Show that

$$\cos(nx) = T_n(\cos(x)), \qquad \sin(nx) = \sin(x)U_{n-1}(\cos(x))$$

for every positive integer n.   ◇

Problem 2.6.33. Show that, for every $n \geq 0$,

$$T_n(x) = \sum_k \binom{n}{2k}(x^2 - 1)^k x^{n-2k},$$

$$U_n(x) = \sum_k \binom{n+1}{2k+1}(x^2 - 1)^k x^{n-2k}.$$ ◇

The polynomials $T_n(x)$ are the *Chebyshev polynomials of the first kind* and the polynomials $U_n(x)$ are the *Chebyshev polynomials of the second kind*.
    Now we introduce Stirling numbers.
    Consider the entries of the following triangle:

				1			
			0		1		
		0		−1		1	
	0		2		−3		1
0		−6		11		−6		1
0	245		−50		35		−10		1
0	−120	274		−225		85		−15		1

The numbers in this table are known as *Stirling numbers of the first kind*. There is no standard notation for these. Numbering them as in Pascal's triangle, we denote the entry in the (n, k) position by $_1S_k^n$. With this numbering, they are given by the recursion

$$_1S_k^{n+1} = {_1S_{k-1}^n} - n({_1S_k^n}) \text{ where } {_1S_0^0} = 1 \text{ and } {_1S_k^0} = 0 \text{ for } k \neq 0.$$

Problem 2.6.34. Show that, for every $n \geq 0$,

$$x^{(n)} = \sum_{k=0}^{n} {}_1S_k^n x^k.$$ ◇

Now consider the entries of the following triangle:

$$
\begin{array}{ccccccc}
 & & & 1 & & & \\
 & & 0 & & 1 & & \\
 & & 0 & 1 & 1 & & \\
 & 0 & 1 & 3 & 1 & & \\
 0 & 1 & 7 & 6 & 1 & & \\
0 & 1 & 15 & 25 & 10 & 1 & \\
0 & 1 & 31 & 90 & 65 & 15 & 1
\end{array}
$$

The numbers in this table are known as *Stirling numbers of the second kind*. Again there is no standard notation for these. Numbering them as in Pascal's triangle, we denote the entry in the (n, k) position by ${}_2S_k^n$. With this numbering, they are given by the recursion

$${}_2S_k^{n+1} = {}_2S_{k-1}^n + k({}_2S_k^n) \text{ where } {}_2S_0^0 = 1 \text{ and } {}_2S_k^0 = 0 \text{ for } k \neq 0.$$

Problem 2.6.35. Show that, for every $n \geq 0$,

$$x^n = \sum_{k=0}^{n} {}_2S_k^n x^{(k)}.$$ ◇

Stirling numbers have many other important combinatorial interpretations and applications, but we shall not go into those here.

Here is another triangle:

$$
\begin{array}{ccccccccccc}
 & & & & & 1 & & & & & \\
 & & & & 1 & 1 & 1 & & & & \\
 & & & 1 & 2 & 3 & 2 & 1 & & & \\
 & & 1 & 3 & 6 & 7 & 6 & 3 & 1 & & \\
 & 1 & 4 & 10 & 16 & 19 & 16 & 10 & 4 & 1 & \\
 1 & 5 & 15 & 30 & 45 & 51 & 45 & 30 & 15 & 5 & 1 \\
1 & 6 & 21 & 50 & 90 & 126 & 141 & 126 & 90 & 50 & 21 & 6 & 1
\end{array}
$$

In this triangle, the entry in any position is the sum of the entry directly above it and the entries to the immediate left and right of that one.

We again number the rows beginning with $n = 0$ for the top row, and we number the northeast-southwest diagonals beginning with $k = 0$ for the top left diagonal. We see that the entries in the diagonal $k = 0$ are 1, 1, 1, 1, 1, ... beginning with $n = 0$, and so the entry in the $(n, 0)$ position is $W_0(n)$ where $W_0(x)$ is the polynomial $W_0(x) = 1$ of degree 0. We see that the entries in the diagonal $k = 1$ are 1, 2, 3, 4, 5, ... beginning with $n = 1$, and so the entry in the $(n, 1)$ position is $W_1(n)$ where $W_1(x)$ is the polynomial $W_1(x) = x$ of degree 1. We see that the entries in the diagonal $k = 2$ are 1, 3, 6, 10, 15, ... beginning with $n = 1$, and so the entry in the $(n, 2)$ position is $W_2(n)$ where $W_2(x)$ is the polynomial $W_2(x) = x(x + 1)/2$ of degree 2.

Problem 2.6.36. Find a polynomial $W_3(x)$ of degree 3 such that the entry in the $(n, 3)$ position is $W_3(n)$ for every n.　　　　　◇

Problem 2.6.37. Show that for every $k \geq 0$ there is a polynomial $W_k(x)$ of degree k such that the entry in the (n, k) position is $W_k(n)$ for every n.　　　◇

We conclude with an important inequality in more than one variable.

Problem 2.6.38. Let $w_1 > 0, w_2 > 0, \ldots, w_n > 0$ with $w_1 + w_2 + \ldots + w_n = 1$. Let $a_1, a_2, \ldots, a_n$ be any positive real numbers. Show that

$$a_1^{w_1} a_2^{w_2} \cdots a_n^{w_n} \leq w_1 a_1 + w_2 a_2 + \ldots + w_n a_n$$

with equality only when $a_1 = a_2 = \cdots = a_n$.　　　　　◇

As a particularly important special case of this inequality we may take $w_1 = w_2 = \ldots = w_n$. Then this inequality becomes

$$\sqrt[n]{a_1 a_2 \cdots a_n} \leq (a_1 + a_2 + \ldots + a_n)/n$$

with equality only when $a_1 = a_2 = \cdots = a_n$.

This is the well-known arithmetic-geometric mean inequality.

2.7　Problems and Results in Linear Algebra

Problem 2.7.1. Prove the following theorem:

Theorem 2.7.2. *Any homogeneous system of m linear equations in n unknowns, with $m < n$, has a nontrivial solution.*　　　　　◇

A great virtue of linear algebra is that it reduces many problems to questions of counting. This theorem is the basis for all these counting arguments. We derive some of its most important corollaries.

Corollary 2.7.3. *Let the vector space V be spanned by a set S of m elements. Then any set T of n elements, with $n > m$, is linearly dependent.*

Proof. Let $S = \{v_1, \ldots, v_m\}$ and let $T = \{w_1, \ldots, w_n\}$. We need to show that the equation $c_1 w_1 + \ldots + c_n w_n = 0$ has a nontrivial solution.

Since S spans V, for each $i = 1, \ldots, n$, we have that $w_i = a_{1,i}v_1 + \ldots + a_{m,i}v_m$ for some (not necessarily unique, but that is irrelevant) $a_{1,i}, \ldots, a_{m,i}$. Thus the equation

$$c_1 w_1 + \ldots + c_n w_n = 0$$

becomes the equation

$$c_1(a_{1,1}v_1 + \ldots + a_{m,1}v_m) + \ldots + c_n(a_{1,n}v_1 + \ldots + a_{m,n}v_m) = 0$$

and we may regroup terms to obtain the equation

$$(c_1 a_{1,1} + \ldots c_n a_{1,n})v_1 + \ldots + (c_1 a_{m,1} + \ldots c_n a_{w,n})m_n = 0.$$

This will certainly be true if

$$c_1 a_{1,1} + \ldots c_n a_{1,n} = 0,$$

$$\vdots$$

$$c_1 a_{m,1} + \ldots c_n a_{m,n} = 0.$$

But this is a system of m linear equations in the n unknowns $c_1, \ldots, c_n$ and so has a nontrivial solution. $\square$

Definition 2.7.4. A *basis* B of a vector space V is a set of elements of V that is linearly independent and spans V.

Corollary 2.7.5. *Any two bases of a vector space V have the same number of elements.*

Proof. Suppose we have a basis B_1 with p elements. Consider any other basis B_2. Let it have q elements. We want to show $q = p$.

By the definition of a basis, B_1 is linearly independent and spans V, and B_2 is linearly independent and spans V. In particular:

(a) B_1 spans V and B_2 is linearly independent; and

(b) B_2 spans V and B_1 is linearly independent.

Then, from the last corollary:
By (a) we have that $q \leq p$; and
by (b) we have that $p \leq q$.
Hence $q = p$. $\square$

Definition 2.7.6. The *dimension* of a vector space V is the number of elements in any basis of V. (This is either a nonnegative integer or ∞.)

Corollary 2.7.7. *Let V be a vector space of finite dimension n and let $B = \{v_1, \ldots, v_n\}$ be any set of n elements of V. Then the following are equivalent:*

(a) B is a basis of V.

(b) B spans V.

(c) B is linearly independent.

Proof. By definition, a basis of V both spans V and is linearly independent, so (a) implies (b) and (c).

Next we prove (b) implies (c). Suppose (b) is true. If (c) were false, then some vector v_i in B would be a linear combination of the remaining vectors in B, and so the subset $C = \{v_1, \ldots, v_{i-1}, v_{i+1}, \ldots, v_n\}$ would also span V. Thus B, which has n elements, would be a linearly independent set in the vector space V that is spanned by the set C which has $n - 1$ elements, contradicting the above corollary. This is impossible, so (c) must be true.

Next we prove (c) implies (b). Suppose (c) is true. If (b) were false, then some vector w in V would not be in the span of V, so $C = \{v_1, \ldots, v_n, w\}$ would also be linearly independent. Thus C, which has $n + 1$ elements, would be a linearly independent set in the vector space V that is spanned by the set B which has n elements, contradicting the above corollary. This is impossible, so (b) must be true.

Thus if either of (b) or (c) is true, so is the other. But (b) and (c) together are just (a). Thus either of (b) or (c) implies (a). $\qquad\qquad\qquad\qquad\square$

Problem 2.7.8. Let $S = \{p_i(x) \mid i = 0, \ldots, n\}$ be a set of polynomials with $\deg(f_i(x)) = i$ for each i. Show that S is a basis of P_n, the vector space of all polynomials of degree at most n. $\qquad\qquad\diamond$

Problem 2.7.9. Let A_n be the n-by-n matrix whose entry a_{ij} in the (i, j)th position is given by:

$$a_{ij} = 2 \text{ if } i = j,$$
$$= 1 \text{ if } |i - j| = 1,$$
$$= 0 \text{ if } |i - j| > 1.$$

Thus for small values of n these matrices are given by:

$$A_1 = \begin{bmatrix} 2 \end{bmatrix}, \quad A_2 = \begin{bmatrix} 2 & 1 \\ 1 & 2 \end{bmatrix}, \quad A_3 = \begin{bmatrix} 2 & 1 & 0 \\ 1 & 2 & 0 \\ 0 & 1 & 2 \end{bmatrix}, \quad A_4 = \begin{bmatrix} 2 & 1 & 0 & 0 \\ 1 & 2 & 1 & 0 \\ 0 & 1 & 2 & 1 \\ 0 & 0 & 1 & 2 \end{bmatrix}.$$

Find and prove a formula for $\det(A_n)$. $\qquad\qquad\diamond$

Problem 2.7.10. Let $A_{n,m}$ be the $(n + m)$-by-$(n + m)$ matrix whose entry a_{ij} in the (i, j)th position is given by:

$$a_{ij} = 2 \text{ if } i = j, \ 1 \le i \le n,$$
$$= 0 \text{ if } i = j, \ i = n + 1,$$
$$= -2 \text{ if } i = j, \ n + 2 \le i \le n + m,$$

$$= 1 \text{ if } |i - j| = 1,$$
$$= 0 \text{ if } |i - j| > 1.$$

(Note that the matrix $A_{n,0}$ is the matrix A_n of the previous problem.) Find and prove a formula for $\det(A_{n,m})$. ◇

Problem 2.7.11. Let $\{s_k\}_{k=1,2,\dots}$ be an arbitrary sequence. Let B_n be the n-by-n matrix whose entry b_{ij} in the (i, j)th position is given by $b_{ij} = s_{\min(i,j)}$.

Thus for small values of n these matrices are given by:

$$B_1 = \begin{bmatrix} s_1 \end{bmatrix}, \quad B_2 = \begin{bmatrix} s_1 & s_1 \\ s_1 & s_2 \end{bmatrix}, \quad B_3 = \begin{bmatrix} s_1 & s_1 & s_1 \\ s_1 & s_2 & s_2 \\ s_1 & s_2 & s_3 \end{bmatrix}, \quad B_4 = \begin{bmatrix} s_1 & s_1 & s_1 & s_1 \\ s_1 & s_2 & s_2 & s_2 \\ s_1 & s_2 & s_3 & s_3 \\ s_1 & s_2 & s_3 & s_4 \end{bmatrix}.$$

Find and prove a formula for $\det(B_n)$. ◇

Problem 2.7.12. For fixed a, b, and $d_1, \dots, d_n$, let $C_n = C_n(a, b; d_1, \dots, d_n)$ be the matrix all of whose entries above the diagonal are a, all of whose entries below the diagonal are b, and whose diagonal entries are $d_1, \dots, d_n$.

Thus for small values of n these matrices are given by:

$$C_1 = \begin{bmatrix} d_1 \end{bmatrix}, \quad C_2 = \begin{bmatrix} d_1 & a \\ b & d_2 \end{bmatrix}, \quad C_3 = \begin{bmatrix} d_1 & a & a \\ b & d_2 & a \\ b & b & d_3 \end{bmatrix}, \quad C_4 = \begin{bmatrix} d_1 & a & a & a \\ b & d_2 & a & a \\ b & b & d_3 & a \\ b & b & b & d_4 \end{bmatrix}.$$

Of course, $\det(C_1) = d_1$.

(a) Suppose $d_1 = \dots = d_n = 0$. Show that $\det(C_n) = (-1)^{n-1} ab(a^{n-1} - b^{n-1})/(a - b)$ for $n > 1$, $a \neq b$, and $\det(C_n) = (-1)^{n-1}(n - 1)a^n$ for $n > 1$, $a = b$.

(b) In general, show that $\det(C_n) = (bf(a) - af(b))/(b - a)$ for $n > 1$, $a \neq b$, and $\det(C_n) = f(a) - af'(a)$ for $n > 1$, $a = b$, where $f(x)$ is the function $f(x) = (d_1 - x) \cdots (d_n - x)$. ◇

Problem 2.7.13. Let D_n be the n-by-n matrix whose entry d_{ij} in the (i, j)th position is given by $d_{ij} = \binom{i+j-2}{i-1} = \binom{i+j-2}{j-1}$.

Thus the entries in the first row of D_n are $1, 1, 1, \dots, 1$, the entries in the second row of D_n are $1, 2, 3, \dots, n$, the entries in the third row of D_n are $1, 3, 6, \dots, n(n + 1)/2$, etc.

Show that $\det(D_n) = 1$ for every $n \geq 1$. ◇

Problem 2.7.14. Let $f(x) = x^n + a_{n-1}x^{n-1} + \cdots + a_1 + a_0$ be a monic polynomial of degree $n \geq 1$. The *companion matrix* $C(f(x))$ of $f(x)$ is the n-by-n matrix

$$
C\big(f(x)\big) = \begin{bmatrix}
-a_{n-1} & 1 & 0 & \cdots & 0 \\
-a_{n-2} & 0 & 1 & \cdots & 0 \\
& & \vdots & \ddots & \\
-a_1 & 0 & 0 & \cdots & 1 \\
-a_0 & 0 & 0 & \cdots & 0
\end{bmatrix}.
$$

(The 1's are immediately above the diagonal.)
    Show that

$$
\det(xI - C(f(x))) = f(x).
$$

(Here, as usual, I denotes the identity matrix, of the proper size.)   ◇

Problem 2.7.15. Fix a, b, and D, and let A be the matrix

$$
A = \begin{bmatrix} a & bD \\ b & a \end{bmatrix}.
$$

Let $N = \det(A) = a^2 - b^2D$.
    Let v_0 be the vector $v_0 = \begin{bmatrix} 1 \\ 0 \end{bmatrix}$ and let $v_n = Av_{n-1}$ for $n \geq 1$, so that $v_n = A^n v_0$. Write $v_n = \begin{bmatrix} p_n \\ q_n \end{bmatrix}$. Thus $p_0 = 1$ and $q_0 = 0$, and, since $v_1 = \begin{bmatrix} a \\ b \end{bmatrix}$, $p_1 = a$ and $q_1 = b$.
    (a) Show that

$$
p_n = -Np_{n-2} + 2ap_{n-1} \qquad \text{and} \qquad q_n = -Nq_{n-2} + 2aq_{n-1}
$$

for $n \geq 2$.
    (b) Show that, assuming $b \neq 0$ and $D \neq 0$,

$$
p_n = (-Nq_{n-1} + aq_n)/b \qquad \text{and} \qquad q_n = (-Np_{n-1} + ap_n)/(bD)
$$

for $n \geq 1$.
    (c) Show that

$$
p_n^2 - Dq_n^2 = N^n
$$

for $n \geq 0$.   ◇

Problem 2.7.16. The n-by-n *Hilbert matrix* H_n is the n-by-n matrix whose entry h_{ij} in the (i, j)th position is given by $h_{ij} = 1/(i + j - 1)$. Thus for small values of n these matrices are given by:

$$H_1 = \begin{bmatrix} 1 \end{bmatrix}, \quad H_2 = \begin{bmatrix} 1 & 1/2 \\ 1/2 & 1/3 \end{bmatrix},$$

$$H_3 = \begin{bmatrix} 1 & 1/2 & 1/3 \\ 1/2 & 1/3 & 1/4 \\ 1/3 & 1/4 & 1/5 \end{bmatrix}, \quad H_4 = \begin{bmatrix} 1 & 1/2 & 1/3 & 1/4 \\ 1/2 & 1/3 & 1/4 & 1/5 \\ 1/3 & 1/4 & 1/5 & 1/6 \\ 1/4 & 1/5 & 1/6 & 1/7 \end{bmatrix}.$$

Show that

$$\det(H_n) = \left[\frac{(2n)!}{2^n n!} \prod_{k=0}^{n-1} \binom{2k}{k}^2 \right]^{-1}. \qquad \diamond$$

Problem 2.7.17. Let $\mathbb{F}$ be a field with infinitely many elements and let $z_1, \ldots, z_k$ be vectors in $\mathbb{F}^k$. Suppose that the jth entry of z_j is nonzero, for each $j = 1, \ldots, k$. Show that there are elements $c_1, \ldots, c_k$ of $\mathbb{F}$ such that, if $z = c_1 z_1 + \ldots + c_k z_k$, every entry of z is nonzero. $\qquad \diamond$

Problem 2.7.18. Let A be an n-by-n matrix. Let B be any matrix that is similar to A. Then, as is well known, the matrix B must have the same trace as the matrix A, $\text{tr}(B) = \text{tr}(A)$, where the trace of a square matrix is the sum of its diagonal entries. Of course, if A is a scalar matrix, then the only matrix similar to A is A itself.

Let A be any n-by-n matrix that is not a scalar matrix, and let $d_1, \ldots, d_n$ be any numbers with $d_1 + \ldots + d_n = \text{tr}(A)$. Show that A is similar to some matrix B whose diagonal entries are $d_1, \ldots, d_n$. $\qquad \diamond$

Now we come to some problems about integer vectors and integer matrices, by which we mean vectors and matrices with integer entries.

Problem 2.7.19. Recall we have earlier defined (directed) graphs and pseudographs.

(a) Let G be a finite pseudograph. We number the vertices of G by $1, \ldots, k$, and we let A be the adjacency matrix of G, i.e., A is the matrix whose entry in the (i, j)th position is the number of edges joining vertex i to vertex j.

Show that, for any $n \geq 0$, the entry in the (i, j)th position of A^n is the number of paths of length n joining vertex i to vertex j.

(b) Let G be a finite directed pseudograph. In its adjacency matrix A, the entry in the (i, j)th position is the number of edges from vertex i to vertex j.

Show that, for any $n \geq 0$, the entry in the (i, j)th position of A^n is the number of paths of length n from vertex i to vertex j. $\qquad \diamond$

Observe that any symmetric matrix with nonnegative integer entries is the adjacency matrix of some pseudograph, and that any matrix with nonnegative integer entries is the adjacency matrix of some directed pseudograph.

Problem 2.7.20. Prove the following theorem:

Theorem 2.7.21. *(a) The group* $GL_2(\mathbb{Z})$ *is generated by the matrices:*

$$\begin{bmatrix} 1 & 1 \\ 0 & 1 \end{bmatrix}, \quad \begin{bmatrix} 0 & 1 \\ 1 & 0 \end{bmatrix}.$$

(b) The group $SL_2(\mathbb{Z})$ *is generated by the matrices:*

$$\begin{bmatrix} 1 & 1 \\ 0 & 1 \end{bmatrix}, \quad \begin{bmatrix} 0 & -1 \\ 1 & 0 \end{bmatrix}.$$

More simply, the problem is to show that: (a) Any 2-by-2 integer matrix with determinant ± 1 can be expressed as a word in the given two matrices, and

(b) Any 2-by-2 integer matrix with determinant 1 can be expressed as a word in the given two matrices.

For example,

$$\begin{bmatrix} -301 & -34 \\ -62 & -7 \end{bmatrix} = \begin{bmatrix} 1 & 1 \\ 0 & 1 \end{bmatrix}^5 \begin{bmatrix} 0 & 1 \\ 1 & 0 \end{bmatrix} \begin{bmatrix} 1 & 1 \\ 0 & 1 \end{bmatrix}^{-7} \begin{bmatrix} 0 & 1 \\ 1 & 0 \end{bmatrix} \begin{bmatrix} 1 & 1 \\ 0 & 1 \end{bmatrix}^9 \begin{bmatrix} 0 & 1 \\ 1 & 0 \end{bmatrix}. \qquad \diamond$$

Problem 2.7.22. For any $n \geq 2$, let

$$v = \begin{bmatrix} v_1 \\ v_2 \\ \vdots \\ v_n \end{bmatrix}$$

with $v_1, v_2, \ldots, v_n$ relatively prime integers. Show that there is an n-by-n integer matrix M with determinant 1 whose first column is v. $\diamond$

Problem 2.7.23. Let A be an n-by-n integer matrix and suppose that $\det(A) \equiv 1 \pmod{m}$ for some m. Show there is an integer matrix B with $A \equiv B \pmod{m}$ and $\det(B) = 1$. $\diamond$

Problem 2.7.24. For an n-by-k integer matrix A, we define $S(A)$, the integer span of the columns of A, by

$$S(A) = \{\, v \in \mathbb{Z}^n \mid v = Ax \text{ for some } x \in \mathbb{Z}^k \,\}.$$

Show that for any integer matrix A there is an integer matrix B with $\det(B) = 1$, and an integer diagonal matrix D, such that $S(A) = S(BD)$. $\diamond$

Chapter 3

Theorems

In this chapter, we present the proofs of a small number of important theorems. These proofs all use induction or the pigeonhole principle, but they are too involved to have been posed as problems. We include them to further show the importance and utility of these methods.

3.1 Fermat's Theorem on Sums of Two Squares

Our goal in this section is to prove Fermat's famous theorem that every prime p congruent to 1 modulo 4 can be written as the sum of two integer squares, $p = x^2 + y^2$. For example, $5 = 2^2 + 1^2$, $13 = 3^2 + 2^2$, $17 = 4^2 + 1^2$, $29 = 5^2 + 2^2$, ..., $97 = 9^2 + 4^2$, $101 = 10^2 + 1^2$, ..., $997 = 31^2 + 6^2$, $1009 = 28^2 + 15^2$, ..., $9973 = 82^2 + 57^2$, $10009 = 100^2 + 3^2$, ..., $99989 = 230^2 + 217^2$, $100049 = 232^2 + 215^2$, ..., $999961 = 765^2 + 644^2$.

We call a way of writing p as a sum of two squares a *representation* of p as a sum of two squares.

Furthermore, if we require x and y to be nonnegative, and do not distinguish between the order of x and y (i.e., we regard $p = x^2 + y^2$ and $p = y^2 + x^2$ as being the same representation), then it is also the case that the representation of p is unique.

Fermat did not leave a record of his proof, but wrote that it was by his method of "descent." Following this hint, Euler gave a proof by descent, and we present this proof here. As we shall see, descent is just a form of induction. (The method of mathematical induction was not formalized until the nineteenth century, long after the times of Fermat and Euler.) We shall present a second twentieth century proof, due to Thue, using the pigeonhole principle.

Of course, every prime p is either equal to 2, and $2 = 1^2 + 1^2$, or is congruent to either 1 or 3 modulo 4. It is very easy to rule out the last case.

Lemma 3.1.1. *No integer congruent to 3 modulo 4 is the sum of two integer squares.*

Proof. Let z be any integer. If z is even, $z = 2k$, then $z^2 = 4k^2$ so $z^2 \equiv 0 \pmod{4}$. If z is odd, $z = 2k + 1$, then $z^2 = 4k^2 + 4k + 1$ so $z^2 \equiv 1 \pmod{4}$. Thus if $n = x^2 + y^2$, then, depending on whether x and y are even or odd, $n \equiv 0 + 0$, $0 + 1$, $1 + 0$, $1 + 1 \pmod{4}$, i.e., $n \equiv 0$, 1, or 2 $\pmod{4}$, i.e., n cannot be congruent to 3 modulo 4. □

Before beginning the proof of Fermat's theorem we introduce some notions that are extremely important in themselves.

Lemma 3.1.2. *Let p be an odd prime. Then the congruence $x^2 \equiv 1 \pmod{p}$ has exactly two solutions: $x \equiv 1 \pmod{p}$ and $x \equiv -1 \pmod{p}$.*

Proof. This congruence is equivalent to the congruence $x^2 - 1 \equiv 0 \pmod{p}$ which is equivalent to the condition that p divides $x^2 - 1 = (x - 1)(x + 1)$. Since p is a prime, this can only happen if p divides the first factor, in which case $x \equiv 1 \pmod{p}$ or if p divides the second factor, in which case $x \equiv -1 \pmod{p}$. □

Theorem 3.1.3 (Wilson). *Let p be an odd prime. Then*

$$(p - 1)! \equiv -1 \pmod{p}.$$

Proof.

$$(p - 1)! = 1 \cdot (2 \cdot 3 \cdots (p - 3) \cdot (p - 2)) \cdot (p - 1).$$

Now for any integer $a \not\equiv 0 \pmod{p}$, there is an integer b with $ab \equiv 1 \pmod{p}$. If $2 \leq a \leq p - 2$, then $b \not\equiv a \pmod{p}$ as otherwise we would have $a^2 \equiv 1 \pmod{p}$ with $a \not\equiv \pm 1 \pmod{p}$, contradicting the last lemma. Thus we may group the $p - 3$ integers between 2 and $p - 2$ into $(p - 3)/2$ pairs of integers $\{a, b\}$ with $ab \equiv 1 \pmod{p}$. But then

$$(p - 1)! \equiv 1 \cdot (1)^{(p-3)/2} \cdot (p - 1) \equiv -1 \pmod{p}. \qquad □$$

Definition 3.1.4. Let p be an odd prime. An integer $a \not\equiv 0 \pmod{p}$ is a *quadratic residue* mod p if there is an integer x with $x^2 \equiv a \pmod{p}$. Otherwise, a is a *quadratic nonresidue* mod p.

Theorem 3.1.5. *Let p be an odd prime. If $p \equiv 1 \pmod{4}$, then -1 is a quadratic residue mod p. If $p \equiv 3 \pmod{4}$, then -1 is a quadratic nonresidue mod p.*

Proof. For any odd prime p, we have, by Wilson's theorem,

$$-1 \equiv (p-1)! \pmod p$$
$$\equiv 1 \cdot 2 \cdots ((p-1)/2) \cdot ((p+1)/2) \cdots (p-2)(p-1) \pmod p$$
$$\equiv 1 \cdot 2 \cdots ((p-1)/2) \cdot (p-(p-1)/2) \cdots (p-2)(p-1) \pmod p$$
$$\equiv 1 \cdot 2 \cdots ((p-1)/2) \cdot (-(p-1)/2) \cdots (-2)(-1) \pmod p$$
$$\equiv (1 \cdot 2 \cdots ((p-1)/2))^2 (-1)^{(p-1)/2} \pmod p$$

as there are $(p-1)/2$ minus signs.

Now suppose $p \equiv 1 \pmod 4$. Then $(p-1)/2$ is even, so $(-1)^{(p-1)/2} = 1$, and so

$$-1 \equiv (1 \cdot 2 \cdots ((p-1)/2))^2 = (((p-1)/2)!)^2 \pmod p$$

and, in particular, -1 is a quadratic residue mod p.

Also, for any odd prime p, if -1 is a quadratic residue mod p, $-1 \equiv x^2 \pmod p$, then $(-1)^{(p-1)/2} \equiv (x^2)^{(p-1)/2} \equiv x^{(p-1)} \equiv 1 \pmod p$ by Fermat's little theorem.

Now suppose $p \equiv 3 \pmod 4$. Then $(p-1)/2$ is odd, so $(-1)^{(p-1)/2} = -1$. Thus if -1 were a quadratic residue mod p we would have $-1 \equiv 1 \pmod p$, which is impossible. $\square$

Corollary 3.1.6. *Let p be an odd prime.*

If $p \equiv 1 \pmod 4$, then for any a there exists a b such that $a^2 + b^2 \equiv 0 \pmod p$.

If $p \equiv 3 \pmod 4$, then $a^2 + b^2 \equiv 0 \pmod p$ if and only if $a \equiv b \equiv 0 \pmod p$.

Proof. If $p \equiv 1 \pmod 4$, then for any a set $b = a((p-1)/2)!$ and then $a^2 + b^2 \equiv a^2 + a^2(-1) \equiv 0 \pmod p$.

Now let $p \equiv 3 \pmod 4$ and suppose that $a^2 + b^2 \equiv 0 \pmod p$. If $a \not\equiv 0 \pmod p$, let c be an integer with $ac \equiv 1 \pmod p$. Then $0 \equiv (a^2 + b^2)c^2 \equiv (ac)^2 + (bc)^2 \equiv 1 + (bc)^2 \pmod p$, and so $(bc)^2 \equiv -1 \pmod p$, which is impossible as -1 is a quadratic nonresidue mod p. Thus we must have $a \equiv 0 \pmod p$, and then, since $a^2 + b^2 \equiv 0 \pmod p$, $b \equiv 0 \pmod p$ as well. $\square$

With these preliminaries out of the way, we get down to proving the theorem. We begin with the following easy algebraic lemma.

Lemma 3.1.7. *Let $m = a^2 + b^2$ and $n = c^2 = d^2$. Then*

$$mn = (a^2 + b^2)(c^2 + d^2) = |ac - bd|^2 + |bc + ad|^2$$
$$= |ac + bd|^2 + |bc - ad|^2.$$

Definition 3.1.8. The two representations of mn as a sum of two squares given by the above lemma are the *compositions* of the given representations of m and n.

Proof. Direct computation. $\square$

Theorem 3.1.9 (Fermat). *Let p be a prime with p not congruent to $3 \bmod 4$. Then $p = x^2 + y^2$ for some nonnegative integers x and y, unique up to order.*

First proof. (Fermat-Euler) We first concentrate on proving the existence part of the theorem, i.e., that every such prime p has a representation as a sum of two squares, and then prove uniqueness.

Suppose that p is not congruent to $3 \bmod 4$. There are two possibilities: $p = 2$ or p congruent to $1 \bmod 4$. If $p = 2$ the theorem is trivial (both existence and uniqueness) as $2 = 1^2 + 1^2$ and this is the only possible representation. Thus we now assume that p is congruent to $1 \bmod 4$.

Let $p \equiv 1 \pmod 4$ and consider *any* pair of integers a and b with $a^2 + b^2 \equiv 0 \pmod p$ and with $a \not\equiv 0 \pmod p$ and $b \not\equiv 0 \pmod p$. We know that at least one such pair exists by our lemma above (for example, $a = 1$ and $b = ((p-1)/2)!$).

Since we are dealing with congruences, we may assume that $1 \le a \le p - 1$ and $1 \le b \le p - 1$. Replacing a by $p - a$ and b by $p - b$ if necessary, we may assume that $1 \le a \le (p-1)/2$ and $1 \le b \le (p-1)/2$. Dividing by $\gcd(a, b)$ if necessary, we may assume that a and b are relatively prime.

Hence $N = a^2 + b^2 < p^2/2$, so $N = pN'$ with $N' < p/2$. In particular, all prime divisors of N' are less than p.

Let p' be any prime divisor of N'. We claim that $p' \not\equiv 3 \pmod 4$. To see this, suppose that $p' \equiv 3 \pmod 4$. Then $a^2 + b^2 = N$ and N is a multiple of p', so $a^2 + b^2 \equiv 0 \pmod{p'}$, and then by our lemma $a \equiv 0 \pmod{p'}$ and $b \equiv 0 \pmod{p'}$, i.e, both a and b are divisible by p', which contradicts our hypothesis that a and b are relatively prime. Hence we must have $p' = 2$ or $p' \equiv 1 \pmod 4$.

Lemma 3.1.10. *Let N be any integer and suppose that $N = a^2 + b^2$ with a and b both nonnegative integers. Let q be any prime divisor of N and suppose that $q = x^2 + y^2$ with x and y nonnegative integers. Then $N/q = u^2 + v^2$ with u and v nonnegative integers such that the representation $N = a^2 + b^2$ is obtained from the representations $N/q = u^2 + v^2$ and $q = x^2 + y^2$ by composition.*

Proof of lemma. Instead of looking for a representation of N/q, we first look at representations of Nq.

By composition, Nq has the representations

$$Nq = |ax + by|^2 + |ay - bx|^2$$
$$= |ax - by|^2 + |ay + bx|^2. \tag{$*$}$$

Now q divides N so

$$q \text{ divides } Ny^2 - b^2q = (a^2 + b^2)y^2 - b^2q = a^2y^2 - b^2(q - y^2)$$
$$= a^2y^2 - b^2x^2 = (ay - bx)(ay + bx).$$

Hence, since q is a prime, q divides one of the factors, i.e., q divides $(ay - bx)$ or q divides $(ay + bx)$. Whichever is the case, choose the corresponding equation in $(*)$.

Then q divides the left-hand side and q divides the second term on the right-hand side, so q divides the first term on the right-hand side as well. Hence

$$ax + by = qu \qquad \text{or} \qquad ax - by = q(\pm u)$$
$$ay - bx = q(\pm v) \qquad\qquad ax - by = qv. \qquad (**)$$

Then, by $(*)$,

$$Nq = (qu)^2 + (qv)^2 \quad \text{and so} \quad N/q = u^2 + v^2$$

and so we have obtained a representation of N/q as a sum of two squares. Furthermore, solving $(**)$ for a and b shows that the representation $N = a^2 + b^2$ is a composition of this representation and the representation $q = x^2 + y^2$. (There are four possibilities. The first possibility is that we have the pair of equations on the left with a sign of $+v$, and in that case we obtain $a = ux + vy$ and $b = uv - vx$. The other cases are similar.) □

Completion of proof of existence, by descent. As we have observed, there is some N that has a representation as a sum of two squares with N a multiple of p, $N = pN'$, and with every prime p' dividing N' not congruent to 3 mod 4 and with $p' < p$. If every such prime had a representation as a sum of two squares, then we could successively apply the above lemma to obtain a representation of p as a sum of two squares. Thus, if p *does not* have a representation as a sum of two squares, there must be a $p' < p$, with p' not congruent to 3 mod 4 that also *does not*. Applying the lemma again, there must be a prime $p'' < p'$, with p'' not congruent to 3 mod 4 that also *does not*. Applying the lemma successively, we obtain a strictly decreasing sequence

$$p > p' > p'' > p''' > \dots$$

of primes, all not congruent to 3 mod 4, *none of which* has a representation as a sum of two squares. But such a sequence must be finite. (This was one of our first consequences of mathematical induction. While mathematical induction was not formally formulated until the nineteenth century, long after the time of Fermat and Euler, they both knew this.) Hence after finitely many steps we arrive at the smallest prime not congruent to 3 mod 4, which is $p = 2$. But $p = 2$ evidently *does* have a representation as a sum of two squares, $2 = 1^2 + 1^2$, so this is impossible, completing the proof. □

Instead of using descent, let us use induction more directly.

Completion of proof of existence, by induction. Let p be any prime not congruent to 3 mod 4. We have as our inductive hypothesis that every prime $p' < p$ with p not congruent to 3 mod 4 has a representation as a sum of two squares, and we wish to show that p does as well.

To begin the induction we observe that 2 has a representation as a sum of two squares, $2 = 1^2 + 1^2$.

Now suppose the inductive hypothesis is true and consider the prime p not congruent to 3 mod 4. Then, as we have shown, there is some N with N of the form $N = pp_1p_2 \cdots p_k$ with each p_i less than p and with each p_i not congruent to 3 mod 4. By the inductive hypothesis, each p_i has a representation as a sum of two squares. We apply the lemma successively to conclude that N/p_k, $N/(p_k p_{k-1})$, ..., and finally $N/(p_k p_{k-1} \cdots p_1) = p$ has a representation as a sum of two squares.

Then, by induction, every such prime p has a representation.   □

Proof of uniqueness. Let p be a prime congruent to 1 mod 4 and suppose p has the representations $p = a^2 + b^2$ and $p = c^2 + d^2$. In the above lemma let $N = p = c^2 + d^2$ and let $q = p = a^2 + b^2$ to conclude that N/q has a representation $N/q = u^2 + v^2$ such that $c^2 + d^2$ is obtained from $u^2 + v^2$ and $a^2 + b^2$ by composition. But $N/q = p/p = 1$ so $u^2 + v^2 = 1$ which forces $u = 1$ and $v = 0$ or vice-versa, so $c = a$ and $d = b$ or $d = a$ and $c = b$.   □

Second proof. (Thue) We begin with a lemma.

Lemma 3.1.11. *Let p be a prime and let a be an integer not divisible by p. Then the congruence $ax \equiv y \pmod{p}$ has a solution x_0, y_0 with $0 < |x_0| < \sqrt{p}$ and $0 < |y_0| < \sqrt{p}$.*

Proof of lemma. Let $k = [\sqrt{p}] + 1$. Note that $k^2 > p$.
     Consider

$$S = \{ax - y \mid 0 \le x \le k - 1, \ 0 \le y \le k - 1\}.$$

There are k^2 possible values for elements of S and $p < k^2$ congruence classes mod p so by the pigeonhole principle we must have two distinct elements of S congruent mod p,

$$ax_1 - y_1 \equiv ax_2 - y_2 \pmod{p} \quad \text{with } (x_1, y_1) \ne (x_2, y_2).$$

But note that we must have both $x_1 \ne x_2$ and $y_1 \ne y_2$ (as either pair being equal forces the other pair to be equal). But then

$$a(x_1 - x_2) \equiv y_1 - y_2 \pmod{p}, \text{ i.e., } ax_0 \equiv y_0 \pmod{p},$$
$$\text{where } \quad x_0 = x_1 - x_2, \ y_0 = y_1 - y_2,$$

and note that since $0 \le x_1, x_2 < k$ and since $0 \le y_1, y_2 < k$ we have that $0 < |x_0| < \sqrt{p}$ and $0 < |y_0| < \sqrt{p}$.   □

Proof of theorem. First we prove existence.
     If $p = 2$, then $2 = 1^2 + 1^2$. Otherwise $p \equiv 1 \pmod{4}$, so -1 is a quadratic residue mod p. Choose a with $a^2 \equiv -1 \pmod{p}$ and let $ax_0 \equiv y_0 \pmod{p}$ as in

the conclusion of the lemma. Then

$$(ax_0)^2 \equiv y_0^2 \pmod{p}$$

$$-x_0^2 \equiv y_0^2 \pmod{p}$$

$$0 \equiv x_0^2 + y_0^2 \pmod{p}$$

so $x_0^2 + y_0^2 = kp$ for some k. But $0 < x_0^2 < p$ and $0 < y_0^2 < p$ so $0 < x_0^2 + y_0^2 < 2p$. Hence we must have $x_0^2 + y_0^2 = p$. Finally, if $x_0 < 0$ we replace x_0 by $-x_0$, and if $y_0 < 0$ we replace y_0 by $-y_0$.

Next we prove uniqueness.

Suppose $p = a^2 + b^2 = c^2 + d^2$ with a, b, c, d all positive. Then

$$a^2 d^2 - b^2 c^2 = a^2 d^2 - b^2 d^2 + b^2 d^2 - b^2 c^2$$

$$= (a^2 + b^2)d^2 - (c^2 + d^2)b^2$$

$$= pd^2 - pb^2 = p(d^2 - b^2) \equiv 0 \pmod{p}.$$

Hence $a^2 d^2 \equiv b^2 c^2 \pmod{p}$, i.e., $(ad)^2 \equiv (bc)^2 \pmod{p}$, and so p divides $(ad)^2 - (bc)^2 = (ad - bc)(ad + bc)$. Since p is a prime, it must divide one of these factors. Hence $ad \equiv bc \pmod{p}$ or $ad \equiv -bc \pmod{p}$. Since $a, b, c, d < \sqrt{p}$ we must have either I: $ad - bc = 0$ or II: $ad + bc = ap$. In case I, $ad = bc$. In case II, $p^2 = (a^2 + b^2)(c^2 + d^2) = (ad + bc)^2 = (ac - bd)^2 = p^2 + (ac - bd)^2$, so $ac = bd$.

Hence either $ad = bc$ or $ac = bd$. Let us assume $ad = bc$. (Otherwise, switch c and d.) Then a divides bc. But a and b are relatively prime (as if they had a common divisor e, e^2 would divide $a^2 + b^2 = p$) so by Euclid's lemma, a divides c. Let $c = ka$. Then $ad = bc = b(ka)$ gives $d = kb$ and then $p = c^2 = d^2 = (ka)^2 + (kb)^2 = k^2(a^2 + b^2) = k^2 p$ so $k = 1$, and then $c = a$ and $d = b$, so the two representations are the same.   □

Given this theorem, it is not hard to see exactly which integers can be written as a sum of two squares, and Fermat knew this as well. (We use the standard language that an integer is squarefree if it is not divisible by any perfect square other than 1.)

Corollary 3.1.12. *A positive integer n can be expressed as $n = x^2 + y^2$ for some integers x and y if and only if $n = N^2 m$ for some integers N and m, with m squarefree and m having no prime factor congruent to 3 mod 4.*

Proof. First suppose that n is of this form. Then we may factor m as a product of primes, $m = p_1 p_2 \cdots p_k$ with each p_i either equal to 2 or congruent to 1 mod 4. By Fermat's theorem, each p_i has a representation as a sum of two squares. Then, applying our algebraic lemma repeatedly, m has such a representation, $m = a^2 + b^2$ for some integers a and b. But then $n = (Na)^2 + (Nb)^2$.

We further claim that only integers of this form have a representation as a sum of two squares. Suppose that claim is false. Then (by well-ordering) there is a smallest integer n not of this form with $n = a^2 + b^2$ for some integers a and b. Thus $n = N^2 m$ where m has some prime factor p congruent to 3 mod 4.

Now $a^2 + b^2 = n = N^2 m$ and p divides m, so in particular $a^2 + b^2 \equiv 0 \pmod{p}$. As we have already seen, since p is congruent to 3 mod 4, we must have that p divides a and p divides b. Let $a = pc$ and $b = pd$. Then $n = N^2 m = p^2(c^2 + d^2)$. Now p^2 does not divide m (since by hypothesis m is squarefree), so we must have that p divides N. Hence $n' = n/p^2 = (N/p)^2 m = c^2 + d^2$ has a representation as a sum of two squares, and $n' < n$, contradicting the minimality of n.   □

Corollary 3.1.13. *A positive integer n can be expressed as $n = x^2 + y^2$ for some integers x and y if and only if, for every prime $p \equiv 3 \pmod{4}$ that divides n, the highest power of p dividing n is even.*

Proof. It is easy to check that the hypothesis of this corollary is equivalent to the hypothesis of the last corollary.   □

3.2 Solutions to Pell's Equation

Let D be a positive integer. We consider the equation $x^2 - Dy^2 = 1$ and look for solutions in integers x and y. This equation always has the trivial solutions $x = \pm 1$, $y = 0$. If D is a perfect square, say $D = d^2$, then $1 = x^2 - Dy^2 = x^2 - d^2 y^2 = (x + dy)(x - dy)$ implies $x + dy = x - dy = 1$ or $x + dy = x - dy = -1$ and so we see that this equation only has the trivial solutions.

Thus we assume henceforth that D is a positive integer that is not a perfect square and we consider the equation $x^2 - Dy^2 = 1$, with x and y integers. This equation is known as *Pell's equation*.

A priori, there is no reason to expect that this equation has a nontrivial solution. Experimentation shows that for small values of D, it does. For example:

For $D = 2$, the smallest solution to Pell's equation in positive integers is $x = 3$, $y = 2$.

For $D = 3$, the smallest solution to Pell's equation in positive integers is $x = 2$, $y = 1$.

For $D = 5$, the smallest solution to Pell's equation in positive integers is $x = 9$, $y = 4$.

For $D = 6$, the smallest solution to Pell's equation in positive integers is $x = 5$, $y = 2$.

For $D = 7$, the smallest solution to Pell's equation in positive integers is $x = 8$, $y = 3$.

For $D = 8$, the smallest solution to Pell's equation in positive integers is $x = 3$, $y = 1$.

For $D = 10$, the smallest solution to Pell's equation in positive integers is $x = 19$, $y = 6$.

For $D = 11$, the smallest solution to Pell's equation in positive integers is $x = 10$, $y = 3$.
For $D = 12$, the smallest solution to Pell's equation in positive integers is $x = 7$, $y = 4$.

We might begin to doubt that this is always the case, as a solution for $D = 13$ does not appear so readily. But there is one: For $D = 13$, the smallest solution to Pell's equation in positive integers is $x = 649$, $y = 180$.

However, we might begin to seriously doubt that this is always the case, as a solution for $D = 61$ does not appear readily at all. But there is one: For $D = 61$, the smallest solution to Pell's equation in positive integers is $x = 1766319049$, $y = 226153980$. (This solution was already known to Fermat.)

Actually, we need to define what we mean by "smallest," since a solution consists of a pair of integers. But note that if $x^2 - Dy^2 = 1$, then $x = \sqrt{1 + Dy^2}$ and $y = \sqrt{(x^2 - 1)/D}$. From this we see that if we have two solutions (x, y) and (x', y') in nonnegative integers, then $x' > x$ if and only if $y' > y$. Thus we may order solutions in nonnegative integers by the size of x, or of y; these two orderings will yield the same answer (and the smallest such solution will be the trivial one $(1, 0)$).

Our aim here is to prove that Pell's equation always has a nontrivial solution and, in fact, infinitely many nontrivial solutions.

Theorem 3.2.1. *Let D be a positive integer that is not a perfect square. Then the equation*

$$x^2 - Dy^2 = 1$$

has a nontrivial solution in integers.

Our proof will involve two applications of the pigeonhole principle, one to prove a lemma and the other to prove the theorem itself.

Lemma 3.2.2. *Let s be any positive integer. Then there exist positive integers t and u with $u \leq s$ and $|t - u\sqrt{D}| < 1/s$.*

Proof. For each $q = 0, 1, \ldots, s$, let r be the nonnegative integer defined by $q\sqrt{D} \leq r < q\sqrt{D} + 1$, so that $0 \leq r - q\sqrt{D} < 1$.

Now divide the interval $[0, 1)$ into s subintervals $[0, 1/s), [1/s, 2/s), \ldots, [(s-1)/s, 1)$.

Since we have s subintervals and $s + 1$ choices for q, some subinterval must contain two of these numbers, say $r_1 - q_1\sqrt{D}$ and $r_2 - q_2\sqrt{D}$. Order these so that $q_1 < q_2$ (and hence $r_1 < r_2$). Then, since both of these real numbers lie in an interval of length $1/s$,

$$|(r_2 - q_2\sqrt{D}) - (r_1 - q_1\sqrt{D})| < 1/s$$
$$|(r_2 - r_1) - (q_2 - q_1)\sqrt{D}| < 1/s$$

and so, setting

$$t = r_2 - r_1, \quad \text{and} \quad u = q_2 - q_1$$

we have that $|t - u\sqrt{D}| < 1/s$. $\square$

Corollary 3.2.3. *There exist infinitely many pairs of positive integers (t, u) with* $|t - u\sqrt{D}| < 1/u$.

Proof. Observe that in the above lemma, since $u \leq s$ then also $1/s \leq 1/u$.

Suppose there are only finitely many such pairs $(t_1, u_1), \ldots, (t_n, u_n)$. Let s be any integer with $s > \max(1/|t_1 - u_1\sqrt{D}|, \ldots, 1/|t_n - u_n\sqrt{D}|)$. Apply the above lemma to conclude there is a pair (t_{n+1}, u_{n+1}) with

$$|t_{n+1} - u_{n+1}\sqrt{D}| < 1/s < 1/u_{n+1},$$

a new pair, which is a contradiction.   □

Proof of theorem. Consider an infinite sequence of pairs $(t_1, u_1), (t_2, u_2), \ldots$ as in the corollary. Let (t, u) be any such pair. Then

$$\begin{aligned}
|t^2 - u^2 D| &= |t - u\sqrt{D}||t + u\sqrt{D}| \\
&= |t - u\sqrt{D}||(t - u\sqrt{D}) + 2u\sqrt{D}| \\
&\leq |t - u\sqrt{D}|(|t - u\sqrt{D}| + |2u\sqrt{D}|) \\
&\leq (1/u)(1/u + 2u\sqrt{D}) = (1/u)^2 + 2\sqrt{D} \leq 1 + 2\sqrt{D}.
\end{aligned}$$

Now let m be a nonzero integer with $|m| < 1 + 2\sqrt{D}$, and for each such m, let $j = 0, 1, \ldots, |m| - 1$ and $k = 0, 1, \ldots, |m| - 1$. Note there are only finitely many triples (m, j, k). For each such triple, consider the conditions

$$t^2 - u^2 D = m$$
$$t \equiv j \pmod{m}$$
$$u \equiv k \pmod{m}.$$

Since there are infinitely many pairs, by the pigeonhole principle there must be some condition satisfied by two of these pairs. Call them (v_1, w_1) and (v_2, w_2), and number them so that $w_1 < w_2$, in which case $v_1 < v_2$ as well. Then

$$v_1^2 - w_1^2 D = v_2^2 - w_2^2 D = m$$
$$v_1 \equiv v_2 \pmod{m}$$
$$w_1 \equiv w_2 \pmod{m}$$

and then

$$\begin{aligned}
m^2 &= (v_1^2 - w_1^2 D)(v_2^2 - w_2^2 D) \\
&= ((v_1 - w_1\sqrt{D})(v_1 + w_1\sqrt{D}))((v_2 - w_2\sqrt{D})(v_2 + w_2\sqrt{D})) \\
&= ((v_1 + w_1\sqrt{D})(v_2 - w_2\sqrt{D}))((v_1 - w_1\sqrt{D})(v_2 + w_2\sqrt{D})) \\
&= (X + Y\sqrt{D})(X - Y\sqrt{D})
\end{aligned}$$

where

$$X = v_1 v_2 - w_1 w_2 D$$
$$Y = v_2 w_1 - v_1 w_2.$$

Now m divides Y by the above congruence conditions. Then $m^2 = X^2 - Y^2 D$ so m^2 divides X^2 and hence m divides X. Thus, setting

$$x = |X/m|, \quad \text{and} \quad y = |Y/m|,$$

we have that x and y are positive integers with

$$1 = (x + y\sqrt{D})(x - y\sqrt{D}) = x^2 - y^2 D$$

and so we have obtained a nontrivial solution to Pell's equation. $\square$

Corollary 3.2.4. *Let D be a positive integer that is not a perfect square. Then the equation*

$$x^2 - Dy^2 = 1$$

has infinitely many solutions in integers.

Proof. Examining the above proof, the pigeonhole principle gives us the stronger result that there is some condition satisfied by infinitely many of these pairs. Call them $(v_1, w_1), (v_2, w_2), \ldots$ and number them so that $w_1 < w_2 < \ldots$ and hence $v_1 < v_2 < \ldots$ as well.

Then perform the above construction with (v_1, w_1) and (v_i, w_i) for $i = 2, 3, \ldots$ to obtain infinitely many distinct solutions. $\square$

Remark 3.2.5. Suppose we have a nontrivial solution to Pell's equation. Then $1 = x^2 - Dy^2 = (x + y\sqrt{D})(x - y\sqrt{D})$, and we see from this equation that each of these two factors has the same sign, and one of them has absolute value less than 1, while the other has absolute value greater than 1. In particular:

x is positive and y is positive if and only if $x + y\sqrt{D} > 1$,
x is positive and y is negative if and only if $0 < x + y\sqrt{D} < 1$,
x is negative and y is positive if and only if $-1 < x + y\sqrt{D} < 0$,
x is negative and y is negative if and only if $x + y\sqrt{D} < -1$.

But in fact not only are there always infinitely many solutions to Pell's equation, they have a structure as well.

Corollary 3.2.6. *There is a solution $x_1^2 - Dy_1^2 = 1$ with x_1 and y_1 positive integers that generates all solutions, in the following sense: Write $\alpha = x_1 + y_1\sqrt{D}$. Then $\alpha^n = x_n + y_n\sqrt{D}$ for integers x_n and y_n. If $x^2 - Dy^2 = 1$ for integers x and y, then for some integer n, and $\varepsilon = \pm 1$, $x = \varepsilon x_n$ and $y = \varepsilon y_n$.*

Proof. Let us first consider solutions (x, y) with x and y positive.

Let (x_1, y_1) be the solution with y_1 smallest, or equivalently x_1 smallest. (Such a solution exists by well-ordering.) Set $\alpha = x_1 + y_1\sqrt{D}$. Let (x', y') be another solution, and set $\beta = x' + y'\sqrt{D}$. Now, as a real number, $\alpha > 1$, and $\beta > \alpha$. Thus there is some positive integer n with $\alpha^n \le \beta < \alpha^{n+1}$. Let $\gamma = \beta/\alpha^n$. We claim that $\gamma = 1$, so that $\beta = \alpha^n$.

To see this, we observe

$$
\begin{aligned}
\gamma &= \frac{\beta}{\alpha^n} \\
&= \frac{x' + y'\sqrt{D}}{(x_1 + y_1\sqrt{D})^n} \\
&= \frac{x' + y'\sqrt{D}}{(x_1 + y_1\sqrt{D})^n} \frac{(x_1 - y_1\sqrt{D})^n}{(x_1 - y_1\sqrt{D})^n} \\
&= \frac{(x' + y'\sqrt{D})(x_1 - y_1\sqrt{D})^n}{(x_1^2 - y_1^2 D)^n} \\
&= \frac{(x' + y'\sqrt{D})(x_1 - y_1\sqrt{D})^n}{1^n} \\
&= (x' + y'\sqrt{D})(x_1 - y_1\sqrt{D})^n \\
&= x'' + y''\sqrt{D}
\end{aligned}
$$

for some integers x'' and y''.

It is easy to check that $1 = (x'' + y''\sqrt{D})(x'' - y''\sqrt{D}) = (x'')^2 - (y'')^2 D$, giving a solution to Pell's equation. But $1 \le \gamma < \alpha_1$, which implies that $0 \le y'' < y_1$. But by the minimality of y_1, this implies that $y'' = 0$, and then $x'' = 1$, so $\gamma = 1$ as claimed.

Thus we have shown that any solution (x, y) to Pell's equation with x and y positive integers is given by α^n for some positive integer n. The trivial solution $x = 1$, $y = 0$ is given by $\alpha^0 = 1$. Now consider a solution (x, y) with x positive and y negative. Then $(x, |y|)$ is a solution, so $x + |y|\sqrt{D} = \alpha^n$ for some $n > 0$, in which case $x + y\sqrt{D} = \alpha^{-n}$.

Thus every solution (x, y) with x positive is as claimed, and to obtain the solutions for x negative, we multiply these by -1.   □

Remark 3.2.7. It is evident that the proof of the existence of nontrivial solutions to Pell's equation that we have given in this section, though short, is entirely nonconstructive. In the next section we will see a constructive method for obtaining solutions.

3.3   Two Theorems on Continued Fractions

We continue our development of continued fractions, with our goal being to prove two theorems. The first is Lagrange's theorem and the second shows how to constructively solve Pell's equation.

We begin with Lagrange's theorem. In fact, we shall begin by stating this theorem, and we shall break up the proof into a series of intermediate steps.

Theorem 3.3.1 (Lagrange). *Let x be a quadratic irrationality. Then the continued fraction expansion of x is periodic.*

Definition 3.3.2. Let $y = s + t\sqrt{D}$ be a rational number (if $t = 0$) or a quadratic irrationality in D (if $t \neq 0$). Its *conjugate* $\bar{y}$ is $\bar{y} = s - t\sqrt{D}$.

Lemma 3.3.3. *For any y_1 and y_2, $\overline{y_1 + y_2} = \bar{y}_1 + \bar{y}_2$, $\overline{y_1 - y_2} = \bar{y}_1 - \bar{y}_2$, $\overline{y_1 y_2} = \bar{y}_1 \bar{y}_2$, and $\overline{y_1/y_2} = \bar{y}_1/\bar{y}_2$ assuming $y_2 \neq 0$.*

Proof. Straightforward computation. □

Lemma 3.3.4. *Let x and x' be quadratic irrationalities. Then $x' = \bar{x}$ if and only if x and x' are roots of the same quadratic equation with rational coefficients.*

Proof. Straightforward computation. □

Definition 3.3.5. A quadratic irrationality x is *reduced* if $x > 1$ and $-1 < \bar{x} < 0$.

Lemma 3.3.6. *Let x be reduced. Then x_n is reduced for every n.*

Proof. By induction on n. By assumption, x_0 is reduced. Assume that x_n is reduced. By definition, $x_{n+1} = 1/(x_n - a_n)$ so $x_{n+1} > 1$. Also, $\overline{x_{n+1}} = 1/(\overline{x_n} - a_n)$. Now $\overline{x_n} < 0$ and $a_n \geq 1$, so $\overline{x_n} - a_0 < -1$ and hence $-1 < 1/(\overline{x_n} - a_n) < 0$. Thus x_{n+1} is reduced. □

Lemma 3.3.7. *Let x be any quadratic irrationality. Then there is some n_0 such that x_n is reduced for all $n \geq n_0$.*

Proof of lemma. We know that, for any $n \geq 0$,

$$x = \frac{x_n p_{n-1} + p_{n-2}}{x_n q_{n-1} + q_{n-2}}$$

so

$$\bar{x} = \frac{\overline{x_n} p_{n-1} + p_{n-2}}{\overline{x_n} q_{n-1} + q_{n-2}}$$

and then elementary algebra yields

$$\overline{x_n} = -\frac{q_{n-2}}{q_{n-1}} \cdot \frac{\bar{x} - p_{n-2}/q_{n-2}}{\bar{x} - p_{n-1}/q_{n-1}}.$$

We have shown that $\lim_{n \to \infty} p_n/q_n = x \neq \bar{x}$, so there is some $n_1 \geq 1$ such that the second fraction on the right-hand side is positive for $n = n_1$. Hence $\overline{x_{n_1}} < 0$. But by definition, $x_{n_1+1} = 1/(x_{n_1} - a_{n_1})$ and $a_{n_1} \geq 1$, so $-1 < \overline{x_{n_1+1}} = 1/(\overline{x_{n_1}} - a_{n_1}) < 0$. Of course, $x_n \geq 1$ for every $n \geq 1$, so in particular $x_{n_1+1} > 1$. Thus x_{n_1+1} is reduced, and then by the above lemma x_n is reduced for every $n \geq n_0 = n_1 + 1$. □

Lemma 3.3.8. *Let x be a quadratic irrationality. The complete quotients x_n in the partial fraction expansion of x are all of the form*

$$x_n = \frac{u_n + \sqrt{E}}{v_n}$$

where u_n and v_n are integers satisfying the recursion

$$u_{n+1} = a_n v_n - u_n, \qquad v_{n+1} = (E - u_{n+1}^2)/v_n$$

for every $n \geq 0$.

Proof. We prove that x_n is of the desired form by induction on n, including in the inductive hypothesis the claim that v_n divides $E - u_n^2$.

Of course, $x_0 = x$. Let $x = s + t\sqrt{D}$. Then $(x - s)^2 - t^2 D = 0$, so x is a root of a quadratic equation with rational coefficients. We may clear denominators to obtain a quadratic equation with relatively prime integer coefficients having x as a root. Let that equation be $ax^2 + bx + c = 0$. Since x is one of the two roots of this equation, we have that $x = (-b + \varepsilon\sqrt{b^2 - 4ac})/(2a)$ where $\varepsilon = \pm 1$. There are four cases, depending on the parity of b and the value of ε. Suppose b is odd. We let $E = b^2 - 4ac$. If $\varepsilon = 1$, we let $u_0 = -b$ and $v_0 = 2a$. If $\varepsilon = -1$, we let $u_0 = b$ and $v_0 = -2a$. Suppose b is even. We let $E = (b/2)^2 - ac$. If $\varepsilon = 1$, we let $u_0 = -(b/2)$ and $v_0 = a$. If $\varepsilon = -1$, we let $u_0 = (b/2)$ and $v_0 = -a$. We observe that in all cases, v_0 divides $E - u_0^2$. We also observe that there is no integer greater than 1 that is a common divisor of u_0 and v_0 and whose square is a divisor of E.

Now assume $x_n = (u_n + \sqrt{D})/v_n$ and also that v_n divides $E - u_n^2$, and consider x_{n+1}. By definition, $x_{n+1} = 1/(x_n - a_n)$. Then, using the induction hypothesis, and doing some elementary (but tricky) algebra, we have

$$x_n = \frac{1}{x_n - a_n}$$

$$= \frac{1}{(u_n + \sqrt{E})/v_n - a_n}$$

$$= \frac{v_n}{(u_n - a_n v_n) + \sqrt{E}}$$

$$= \frac{v_n(u_n - a_n v_n) - \sqrt{E}}{(u_n - a_n v_n)^2 - E}$$

$$= \frac{(a_n v_n - u_n) + \sqrt{E}}{(E - (a_n v_n - u_n)^2)/v_n}$$

and so we obtain the expressions in the lemma. It remains to be shown that u_{n+1} and v_{n+1} are integers and that v_{n+1} divides $E - u_{n+1}^2$. Clearly u_{n+1} is an integer. By hypothesis, v_n divides $E - u_n^2$ so v_n divides $E - (a_n v_n - u_n)^2$ and v_{n+1} is an integer. Finally, $v_n v_{n+1} = E - u_{n+1}^2$ with v_n an integer, so v_{n+1} divides $E - u_{n+1}^2$. Thus x_{n+1} has an expression as claimed and by induction we are done. $\square$

Remark 3.3.9. Note that if $x = \sqrt{D}$ for an integer D that is not a perfect square, the above construction yields $E = D$, $u_0 = 0$, and $v_0 = 1$, i.e., it is just the expression $\sqrt{D} = (0 + \sqrt{D})/1$.

Conclusion of proof of theorem. Choose n_0 so that x_n is reduced for $n \geq n_0$. Then

$$\frac{u_n + \sqrt{E}}{v_n} > 1 \quad \text{and} \quad -1 < \frac{u_n - \sqrt{E}}{v_n} < 0$$

for $n \geq n_0$. Subtracting, we see that $2\sqrt{E}/v_n > 0$, so $v_n > 0$. But $v_n > 0$ gives $u_n + \sqrt{E} > 0$ and $u_n - \sqrt{E} < 0$ so $-\sqrt{E} < u_n < \sqrt{E}$. But $u_n + \sqrt{E} > v_n$ so $v_n < 2\sqrt{E}$. Summarizing, we have the inequalities

$$-\sqrt{E} < u_n < \sqrt{E} \quad \text{and} \quad 0 < v_n < 2\sqrt{E}$$

for $n \geq n_0$. But u_n and v_n are integers. Thus there are only finitely many possibilities for the pair (u_n, v_n) and so, by the pigeonhole principle, there must be two integers, which we write as k and $k + m$ with $m > 0$, such that $u_{k+m} = u_k$ and $v_{k+m} = v_k$, and hence $x_{k+m} = x_k$. But that implies $a_{k+m} = a_k$; then $x_{k+1+m} = 1/(x_{k+m} - a_{k+m}) = 1/(x_k - a_k) = x_{k+1}$ so $a_{k+1+m} = a_{k+1}$, and continuing (or, more precisely, arguing inductively), $a_{j+m} = a_j$ for every $j \geq k$, and the continued fraction expansion of x is periodic. $\qquad\square$

Now for solving Pell's equation.

We begin with two results that are interesting in their own right.

Lemma 3.3.10. *Suppose that $x = \sqrt{D}$ where D is an integer that is not a perfect square. Then, in the above notation,*

$$p_{n-1}^2 - Dq_{n-1}^2 = (-1)^n v_n$$

for every $n \geq 0$.

Proof. We know that

$$x = \frac{x_n p_{n-1} + p_{n-2}}{x_n q_{n-1} + q_{n-2}}$$

$$\sqrt{D} = \frac{(u_n + \sqrt{D}/v_n) p_{n-1} + p_{n-2}}{(u_n + \sqrt{D}/v_n) q_{n-1} + q_{n-2}}$$

and then elementary algebra yields

$$(p_{n-1} u_n + p_{n-2} v_n) + p_{n-1}\sqrt{D} = q_{n-1} D + (q_{n-1} u_n + q_{n-2} v_n)\sqrt{D}$$

and so

$$q_{n-1} u_n + q_{n-2} v_n = p_{n-1} \quad \text{and} \quad p_{n-1} u_n + p_{n-2} v_n = q_{n-1} D.$$

Multiplying the first of these equations by p_{n-1}, the second by q_{n-1}, and subtracting, we obtain

$$p_{n-1}^2 - Dq_{n-1}^2 = (p_{n-1}q_{n-2} - p_{n-2}q_{n-1})v_n = (-1)^n v_n$$

as claimed.

□

Lemma 3.3.11. *A quadratic irrationality* x *has a purely periodic continued fraction expansion if and only if* x *is reduced.*

Proof. First suppose that x is purely periodic of period m. Then $x = x_0 = x_{tm}$ for every $t \geq 1$. We know that there is an n_0 such that x_n is reduced for every $n \geq n_0$. Choose t so that $tm \geq n_0$. Then $x = x_{tm}$ is reduced.

On the other hand, suppose that x is reduced, and let x have continued fraction expansion $x = [a_0, a_1, \ldots]$ periodic of period m beginning at a_k. Set $z = x_k = x_{m+k}$. Then

$$x_{k-1} = a_{k-1} + 1/z, \quad x_{m+k-1} = a_{m+k-1} + 1/z$$

and so

$$\overline{x_{k-1}} = a_{k-1} + 1/\overline{z}, \quad \overline{x_{m+k-1}} = a_{m+k-1} + 1/\overline{z}.$$

Subtracting,

$$\overline{x_{k-1}} - \overline{x_{m+k-1}} = a_{k-1} - a_{m+k-1}.$$

But each x_n is reduced, so $-1 < \overline{x_n} < 0$ for every n, and that implies

$$-1 < \overline{x_{k-1}} - \overline{x_{m+k-1}} = a_{k-1} - a_{m+k-1} < 1.$$

But a_{k-1} and a_{m+k-1} are both integers, so we must have $a_{k-1} = a_{m+k-1}$ and $\overline{x_{k-1}} = \overline{x_{m+k-1}}$, and hence $x_{k-1} = x_{m+k-1}$. Proceeding by downward induction we find $a_{k-j} = a_{m+k-j}$ for all $j \leq k$ or, equivalently, $a_j = a_{j+m}$ for all $j \geq 0$, i.e., the continued fraction expansion for x is purely periodic. □

We also need the following result.

Lemma 3.3.12. *Let* y *be any real number, and let* $y' = y + e$ *for any integer* e. *Let* y *have continued fraction expansion* $y = [a_0, a_1, \ldots]$ *with convergents* p_n/q_n *and let* $y_n = (u_n + \sqrt{D})/v_n$ *as above. Let* y' *have continued fraction expansion* $y = [a_0', a_1', \ldots]$ *with convergents* p_n'/q_n' *and let* $y_n' = (u_n' + \sqrt{D'})/v_n'$ *as above. Then* $a_0' = a_0 + e$ *and* $a_n' = a_n$ *for every* $n \geq 1$. *Also,* $p_{-2}' = p_{-2}$, $q_{-2}' = q_{-2}$, $p_{-1}' = p_{-1}$, $q_{-1}' = q_{-1}$, *and* $p_n' = p_n + eq_n$, $q_n' = q_n$ *for every* $n \geq 0$. *Also,* $D' = D$, $u_0' = u_0 + ev_0$, $v_0' = v_0$, *and* $u_n' = u_n$, $v_n' = v_n$ *for every* $n \geq 1$.

Proof. Straightforward.

□

Theorem 3.3.13. *Let* D *be a positive integer that is not a perfect square. Let* $C_n = p_n/q_n$, $n = 0, 1, \ldots$, *be the convergents of the continued fraction expansion* *of $x = \sqrt{D}$, and let m be the period of this continued fraction expansion. Then for every positive integer t,*

$$p_{tm-1}^2 - Dq_{tm-1}^2 = (-1)^m.$$

In particular,

> *If m is even, $p_{tm-1}^2 - Dq_{tm-1}^2 = 1$ for every t;*
>
> *If m is odd, $p_{tm-1}^2 - Dq_{tm-1}^2 = 1$ for every even t.*

Proof. Note that $x = x_0 = (u_0 + \sqrt{D})/v_0 = (0 + \sqrt{D})/1$ and $u_0 = 0, v_0 = 1$.

Let d be the unique positive integer with $d^2 < D < (d+1)^2$. Let $x' = d + \sqrt{D}$. Then $\overline{x'} = d - \sqrt{D}$. Thus $x' > 1$ and $-1 < \overline{x'} < 0$, i.e., x' is reduced. Thus x' has a purely periodic continued fraction expansion $[2d, a_1', \ldots, a_{m-1}', 2d, a_1', \ldots]$.

But then x has the periodic continued fraction expansion $[d, a_1, \ldots, a_{m-1}, 2d, a_1, \ldots]$ (periodic beginning with a_1). Thus $p_{tm-1}^2 - Dq_{tm-1}^2 = (-1)^m v_{tm}$. But $v_{tm} = v_{tm}'$. However, x' is purely periodic, so $v_{tm}' = v_0' = v_0 = 1$ and we are done.   □

Example 3.3.14. We shall see how this algorithm works in several cases.

(a) Let $D = 13$. Then the continued fraction expansion of $3 + \sqrt{13}$ is purely periodic, and so the continued fraction expansion of $\sqrt{13}$ is periodic beginning with a_1. We have already computed that $\sqrt{13} = [3, \overline{1,1,1,1,6}]$, periodic of period 5. We have the following table:

n	p_n	q_n	$p_n^2 - 13q_n^2$
-1	1	0	1
0	3	1	-4
1	4	1	3
2	7	2	-3
3	11	3	4
4	18	5	-1
5	119	33	4
6	137	38	-3
7	256	71	3
8	393	109	-4
9	649	180	1
19	842401	233640	1
29	1093435849	303264540	1

(b) Let $D = 19$. Then the continued fraction expansion of $4 + \sqrt{19}$ is purely periodic, and so the continued fraction expansion of $\sqrt{19}$ is periodic beginning with a_1. We have already computed that $\sqrt{19} = [4, \overline{2, 1, 3, 1, 2, 1, 8}]$, periodic of period 6.

We have the following table:

n	p_n	q_n	$p_n^2 - 19q_n^2$
-1	1	0	1
0	4	1	-3
1	9	2	5
2	13	3	-2
3	48	11	5
4	61	14	-3
5	170	39	1
11	57799	13260	1
17	19651490	4508361	1
23	2397964916	550130881	1

(c) Let $D = 61$. Then the continued fraction expansion of $7 + \sqrt{61}$ is purely periodic, and so the continued fraction expansion of $\sqrt{61}$ is periodic beginning with a_1. Computation shows that $\sqrt{61} = [7, \overline{1, 4, 3, 1, 2, 2, 1, 3, 4, 1, 14}]$, periodic of period 11.

We have the following table:

n	p_n	q_n	$p_n^2 - 61q_n^2$
-1	1	0	1
0	7	1	-12
1	8	1	3
2	39	5	-4
3	125	16	9
4	164	21	-5
5	453	58	5
6	1070	137	-9
7	1523	195	4
8	5639	722	-3
9	24079	3083	12
10	29718	3805	-1
11	440131	56352	12
12	469849	60158	-3
13	2319527	296985	4

14	7428430	951113	−9
15	9747957	1248098	5
16	26924344	3447309	−5
17	63596645	8142716	9
18	90520989	11590025	−4
19	335159612	42912791	3
20	1431159437	183241189	−12
21	1766319049	226153980	1
43	5055768351458430833	647324805384218521	1

3.4 A Theorem in Linear Algebra

Theorem 3.4.1. *A vector space V over an infinite field $\mathbb{F}$ is not the union of finitely many proper subspaces.*

We give two proofs of this theorem.

First proof. We prove this in the case that V is finite-dimensional, by induction on the dimension, and then reduce the infinite-dimensional case to the finite-dimensional case.

Let V have dimension n. If $n = 0$, V has no proper subspaces, so that case is empty. If $n = 1$, the only proper subspace of V is $\{0\}$, and $V \neq \{0\}$, so the theorem is true.

Assume the theorem is true for all $(n-1)$-dimensional vector spaces over $\mathbb{F}$, and let V have dimension n. Let V have basis $B = \{v_1, v_2, \ldots, v_n\}$. For every element f of $\mathbb{F}$, we have the subspace V_f of V with basis $\{v_1 + f v_2, v_3, \ldots, v_n\}$. Since $\mathbb{F}$ is infinite, there are infinitely many such subspaces V_f.

Now suppose that V is the union of proper subspaces $W_1, \ldots, W_k$, i.e., $V = W_1 \cup \ldots \cup W_k$. Choose f so that $V_f \neq W_i$, $i = 1, \ldots, k$. Then

$$
\begin{aligned}
V_f = V \cap V_f &= (W_1 \cup \ldots \cup W_k) \cap V_f \\
&= (W_1 \cap V_f) \cup \ldots \cup (W_k \cap V_f) \\
&= U_1 \cup \ldots \cup U_k
\end{aligned}
$$

where $U_i = W_i \cap V_f$, $i = 1, \ldots, k$. But V_f has dimension $n-1$ and each U_i is a proper subspace of V_f, so V_f is a union of finitely many proper subspaces, contradicting the inductive hypothesis. Thus we conclude that V itself is not the union of finitely many proper subspaces, completing the inductive step. Hence, by induction, the theorem is true for all finite-dimensional vector spaces over $\mathbb{F}$.

Now let V be an arbitrary vector space over $\mathbb{F}$, and suppose that V is the union of proper subspaces $W_1, \ldots, W_k$, i.e., $V = W_1 \cup \ldots \cup W_k$. Since each W_i is a proper subspace of V, we may choose v_i in V, v_i not in W_i, $i = 1, \ldots, k$. Let V_0

be the subspace of V spanned by $\{v_1, \ldots, v_k\}$. Since V_0 is spanned by finitely many vectors, it is finite-dimensional (of dimension at most k). But $V = W_1 \cup \ldots \cup W_k$, so (as above)

$$V_0 = (W_1 \cap V_0) \cup \ldots \cup (W_k \cap V_0).$$

Now each $W_i \cap V_0$ is a proper subspace of V_0, as v_i in V_0, v_i not in $W_i \cap V_0$, $i = 1, \ldots, k$. But then the finite-dimensional vector space V_0 is a union of finitely many proper subspaces; contradiction.   □

We prepare for the second proof by proving a general lemma.

Lemma 3.4.2. *Let V be a vector space over an arbitrary field $\mathbb{F}$, and let W be a proper subspace of V. Let v be any element of V that is not in W. Then there is a nonzero linear transformation $T : V \longrightarrow \mathbb{F}$ with $T(v) = 1$ and $T(w) = 0$ for every w in W.*

Proof of lemma. Let $B_W = \{w_i\}$ be a basis of W. Then $B_W \cup \{v\}$ is a linearly independent set, so we may extend it to a basis $B = \{w_i\} \cup \{v\} \cup \{u_j\}$ of V. Then define T by $T(w_i) = 0$ for every i, $T(v) = 1$, and $T(u_j) = 0$ for every j. (Actually, the values $T(u_j)$ are irrelevant, but we have to make a choice, so we chose them to be 0.)   □

Second proof. We prove that V cannot be the union of a finite number of proper subspaces $\{W_1, \ldots, W_k\}$, in the case that $\mathbb{F}$ is infinite, by using this lemma and an earlier result.

For each $i = 1, \ldots, k$, choose a vector v_i in V, v not in W_i, and let $T_i : V \longrightarrow \mathbb{F}$ be a linear transformation with $T_i(w) = 0$ for every w in W and $T_i(v_i) = 1$.

For each $i = 1, \ldots, k$ let z_i be the vector in $\mathbb{F}^k$ defined by

$$z_i = \begin{bmatrix} T_1(v_i) \\ T_2(v_i) \\ \vdots \\ T_k(v_i) \end{bmatrix}$$

so that the jth entry of z_i is $T_j(v_i)$.

Then, by a previous problem, since we are assuming that $\mathbb{F}$ is infinite, there are elements $c_1, \ldots, c_k$ of $\mathbb{F}$ such that every entry of $z = c_1 z_1 + \ldots + c_k z_k$ is nonzero. Let

$$v = c_1 v_1 + \ldots + c_k v_k.$$

Then, for each j, $T_j(v) = T_j(c_1 v_1 + \ldots + c_k v_k) = c_1 T_j(v_1) + \ldots + c_k T_j(v_k)$ is the jth entry of z, and so is nonzero. But $T_j(v) \neq 0$ implies that v is not in W_j. Hence we see that v is an element of V that is not in W_j, for each $j = 1, \ldots, k$, and thus we conclude $W_1 \cup \ldots \cup W_k \neq V$.   □

3.5 Arithmetic Progressions in the Positive Integers

In this section we prove a famous theorem of van der Waerden. The proof, by induction, is entirely elementary but very subtle.

An *arithmetic progression* of length k is a sequence $a_1, a_2, \ldots, a_k$ in which the difference $a_{i+1} - a_i$ is constant for $i = 1, \ldots, k - 1$. If we call this common difference d, then, equivalently, an arithmetic progression of length k is a sequence $a_1, a_1 + d, a_1 + 2d, \ldots, a_1 + (k - 1)d$ for some values of a_1 and d.

Theorem 3.5.1 (van der Waerden). *Let ℓ and k be positive integers. Then there exists an integer $N = N(\ell, k)$ such that if the set $S = \{1, \ldots, N\}$ of integers from 1 to N is partitioned into k subsets, at least one of these subsets must contain an arithmetic progression of length ℓ.*

Proof. It is convenient, and helpful, to think of this in terms of colors. That is, we think of partitioning S by coloring each of the integers 1 through N with one of k colors (so that the first subset consists of those integers colored red, the second subset consists of those colored green, etc.). Then the conclusion becomes that there must be an arithmetic progression of length ℓ all of whose elements have the same color or, in other words, that there must be a monochromatic arithmetic progression of length ℓ.

We prove this by induction on ℓ. For $\ell = 1$ this is completely trivial. For $\ell = 2$ we see that $N(\ell, k) = k + 1$, for any value of k, as, by the pigeonhole principle, if any set S of $k + 1$ elements is colored by k colors, that there must be two elements of the same color (and any sequence of length 2 is automatically an arithmetic progression).

Assume the theorem is true for $\ell - 1$ (and any value of k). Let S be colored by k colors.

We begin by introducing some terminology.

We call an arithmetic progression of length ℓ in S *almost monochromatic* if its first $\ell - 1$ terms all have the same color. Thus an almost monochromatic sequence is or is not monochromatic as its last term does or does not have the same color as all the preceding terms.

We let S_w be any sequence of w consecutive integers in S. We call such a sequence a *w-system*. A *coloration* of a w-system S_w is a way of coloring each of the integers in S_w, and we see that there are k^w possible colorations of a w-system. We say that two w-systems S_w and S'_w have the same coloration if their corresponding elements have the same color. That is, if S_w consists of the integers $a, \ldots, a + (w - 1)$ and S'_w consists of the integers $a', \ldots, a' + (w - 1)$, S_w and S'_w have the same coloration if $a + i$ and $a' + i$ have the same color for each $i = 0, \ldots, w - 1$.

Call two w-systems S_w and S'_w *consecutive* if their starting entries are consecutive, i.e., if $a' = a + 1$.

An *arithmetic progression of w-systems* is a sequence of w-systems $S_w, S'_w, S''_w, \ldots$ whose starting entries $a, a', a'', \ldots$ form an arithmetic progression. We call such an arithmetic progression monochromatic if all of its terms have the same coloration, and almost monochromatic if all of its terms except possibly the last one have the same coloration.

With all this language in place, we get to work.

We observe the theorem is true for S if and only if it is true for any sequence of N consecutive integers. Hence, by the inductive hypothesis, we may assume that in any set of $N(\ell - 1, k^w)$ consecutive w-systems, there is an almost monochromatic arithmetic progression of w-systems of length $\ell - 1$.

Let $\psi(\ell, k)$ be the function

$$\psi(\ell, k) = N(\ell - 1, k) + \left[\frac{N(l - 1, k) - 1}{\ell - 2} \right]$$

where $[\cdot]$ denotes the greatest integer function.

Let $M_1 = \psi(\ell, k)$. We claim that any M_1-system S_{M_1} contains an almost monochromatic arithmetic progression of length ℓ. To see this, we observe that, if $M_0 = N(\ell - 1, k)$, any M_0-system S_{M_0} contains a monochromatic arithmetic progression of length $\ell - 1$. S_{M_0} is a sequence of M_0 positive integers, so the common difference d in the sequence cannot exceed $e = [(M_0 - 1)/(\ell - 2)]$. Hence if we add an additional e consecutive integers to S_{M_0} to obtain S_{M_1}, the resulting sequence of M_1 consecutive integers must contain at least one more term of this arithmetic progression.

By exactly the same logic, we see that if $M_w = \psi(\ell, k^w)$, in any M_w consecutive w-systems in S there must be an almost monochromatic arithmetic progression of w-systems of length ℓ. Recall that a w-system consists of w consecutive integers. Thus, if we let

$$\varphi(w) = \psi(\ell, k^w) + (w - 1),$$

and set $v = \varphi(w)$, we conclude that in any set of v consecutive integers in S, i.e., in any v-system in S, there is an almost monochromatic arithmetic progression of ℓ w-systems.

Set $n_0 = 1$ and $n_{i+1} = \varphi(n_i)$, $i = 1, \ldots, k - 1$. Set

$$N = N(\ell, k) = n_k.$$

The n_k-system S_n then contains an almost monochromatic arithmetic progression of ℓ n_{k-1}-systems

$$S_{n_{k-1}}^0, S_{n_{k-1}}^1, \ldots, S_{n_{k-1}}^{\ell-1},$$

and similarly the n_{k-1}-system $S_{n_{k-1}}^0$ contains an almost monochromatic progression of ℓ n_{k-2}-systems

$$S_{n_{k-2}}^0, S_{n_{k-2}}^1, \ldots, S_{n_{k-2}}^{\ell-1},$$

and so forth, so that for any j between k and 1, the n_j-system S_{n_j} contains an almost monochromatic arithmetic progression of ℓ n_{j-1}-systems

$$S_{n_{j-1}}^0, S_{n_{j-1}}^1, \ldots, S_{n_{j-1}}^{\ell-1}.$$

Let us call this progression P_j. Note that in P_1, the last of these progressions,

$$S_{n_0}^0, S_{n_0}^1, \ldots, S_{n_0}^{\ell-1},$$

each term consists of a single integer, as $n_0 = 1$.

Now we shall show that by a proper choice of integers out of these arithmetic progressions of systems we can obtain an arithmetic progression of integers.

Since P_j is an arithmetic progression of systems, each system in the progression is obtained from the preceding one by adding some positive integer d_j to each term, so for any λ between 0 and $\ell - 1$, $S_{n_{j-1}}^\lambda$ is obtained from $S_{n_{j-1}}^0$ by adding λd_j to each term. In particular, if $S_{n_0}^0$ consists of the single integer a_0, $S_{n_0}^{\lambda_1}$ consists of the single integer $a_0 + \lambda_1 d_1$. Now $S_{n_0}^{\lambda_1}$ is contained in $S_{n_1}^0$ so this integer is certainly an element of $S_{n_1}^0$. By the same logic, $a_0 + \lambda_1 d_1 + \lambda_2 d_2$ is contained in $S_{n_1}^{\lambda_2}$ and hence in $S_{n_2}^0$. Proceeding in this fashion, we see that each of the integers

$$a_0 + \lambda_1 d_1 + \ldots + \lambda_j d_j, \quad 0 \le \lambda_1, \ldots, \lambda_j \le \ell - 1$$

lies in $S_{n_j}^0$.

Now the arithmetic progression P_1 is almost monochromatic, so $a_0 + \lambda_1 d_1$ has the same color as a_0 as long as $\lambda_1 \le \ell - 2$. Similarly, the arithmetic progression P_2 is almost monochromatic, so $a_0 + \lambda_1 d_1 + \lambda_2 d_2$ has the same color as $a_0 + \lambda_1 d_1$ as long as $\lambda_2 \le \ell - 2$. Proceding in this fashion, we see that, for any $p < q$, and for any fixed values of $\lambda_1, \ldots, \lambda_p$, the integers

$$a_0 + \lambda_1 d_1 + \ldots + \lambda_p d_p \text{ and } a_0 + \lambda_1 d_1 + \ldots + \lambda_p d_p + \lambda_{p+1} d_{p+1} + \ldots + \lambda_q d_q$$

have the same color, providing that $0 \le \lambda_{p+1}, \ldots, \lambda_q \le \ell - 2$.

Now consider the $k + 1$ integers

$$a_0, \ a_1 = a_0 + (\ell - 1)d_1, \ldots, \ a_k = a_0 + (\ell - 1)(d_1 + d_2 + \ldots + d_k).$$

There are $k + 1$ of these integers, and only k colors, so two of them must have the same color, say a_p and a_q for $p < q$. Note that $a_q - a_p = (\ell - 1)D$ where $D = d_{p+1} + \ldots + d_q$.

Now consider the arithmetic progression

$$a_p, \ a_p + D, \ a_p + 2D, \ \ldots, \ a_p + (\ell - 2)D, \ a_p + (\ell - 1)D = a_q.$$

The first $\ell - 1$ terms of this arithmetic progression all have the same color, as they are $a_p + \lambda_{p+1} d_{p+1} + \ldots + \lambda_q d_q$ for $\lambda_{p+1} = \ldots = \lambda_q = i$ for $i = 0, \ldots, \ell - 2$. But the last term a_q has the same color as the first term a_p. Thus we have obtained a monochromatic sequence of length ℓ contained in S.

Then by induction we are done. $\square$

Corollary 3.5.2. *Let k be a positive integer. If the set S of positive integers is partitioned into k subsets, at least one of these subsets must contain an arithmetic progression of every length.*

Proof. From the preceding theorem we immediately conclude: Let ℓ and k be positive integers. If the set S of positive integers is partitioned into k subsets, at least one of these subsets must contain an arithmetic progression of length ℓ.

Call these k subsets $T_1, \ldots, T_k$. For each $\ell = 1, 2, \ldots$, there is a subset $T_{j(\ell)}$ containing an arithmetic progression of length ℓ. But there are infinitely many values of ℓ and only finitely many values of $j(\ell)$, so by the pigeonhole principle some value must be taken on infinitely many times. In other words, there is some j_0 such that T_{j_0} contains arithmetic progressions of lengths $\ell_1, \ell_2, \ldots$ with $\ell_1 < \ell_2 < \ldots$, and hence T_{j_0} contains an arithmetic progression of any length.   □

3.6 Public Key Cryptography

Suppose we wish to develop a system where a sender A can transmit an enciphered message to a recipient B over an insecure channel, in such a way that a third party C who intercepts that message cannot decipher it. Suppose further we wish to allow anyone at all to send such a message to B, and so we make the encipherment scheme known to any sender. How can we do this? In other words, how can we develop a system where the encipherment scheme is completely public, but decipherment by anyone except the intended recipient is impossible?

Such a system goes under the general name of *public key cryptography*. In this section we present the basic idea behind *RSA cryptography*, which was the original such system. We will see that this is a very clever application of some of the simple number-theoretic ideas we have already developed. (There are many other considerations that go into actual implementation of RSA, and there have been many other public key systems that have been developed, some involving very sophisticated mathematics, but our point here is not to present a treatise on cryptography; rather, it is to give an interesting application of some of the work we have already done.)

The key to RSA is the following theorem. In this theorem, $\varphi(A)$ is the Euler totient function that we have previously defined.

Theorem 3.6.1. *Suppose A is a product of distinct primes. For any integer b, and for any positive integer t,*

$$b^{t\varphi(A)+1} \equiv b \ (\mathrm{mod} \ A).$$

Proof. First suppose that $A = p$ is a prime. Then $\varphi(p) = p - 1$. If b is relatively prime to p, we have that

$$b^{t(p-1)+1} = \left(b^{p-1}\right)^t b \equiv 1^t b \equiv b \ (\mathrm{mod} \ p)$$

by Fermat's little theorem, while if b is not relatively prime to p, i.e., if p divides b, then certainly

$$b^{t(p-1)+1} \equiv 0^{t(p-1)+1} = 0 \equiv b \ (\mathrm{mod} \ p)$$

so the theorem is true in this case.

Now we consider the general case. First we observe if m and n are any two integers with m dividing n, then $\varphi(m)$ divides $\varphi(n)$. (This follows immediately from the formula we have developed for $\varphi(n)$.)

In particular, if $A = p_1 p_2 \cdots p_k$, then $\varphi(A)$ is divisible by each of $\varphi(p_1), \varphi(p_2), \ldots, \varphi(p_k)$.

Let us now consider the following system of simultaneous congruences.

$$x \equiv b^{t\varphi(A)+1} \pmod{p_1}$$

$$x \equiv b^{t\varphi(A)+1} \pmod{p_2}$$

$$\vdots$$

$$x \equiv b^{t\varphi(A)+1} \pmod{p_k}$$

By the Chinese remainder theorem, we know this system of congruences has a *unique* solution (mod $p_1 p_2 \ldots p_k$), i.e., (mod A). But from the case A a prime we know that each of these congruences has a solution $x \equiv b \pmod{A}$ (as that implies $x \equiv b \pmod{p_i}$ for each i). Since this is *a* solution, and the solution is unique (mod A), it must be *the* solution (mod A), as claimed. $\square$

Remark 3.6.2. If A is not a product of distinct primes, this theorem is false. Suppose that A is divisible by p^2, for some prime p. Let $b = A/p$. Then $b \not\equiv 0$ (mod A) but $b^2 \equiv 0 \pmod{A}$ and so $b^{t(p-1)+1} \equiv 0 \not\equiv b \pmod{A}$.

Here is the basic technique behind RSA cryptography. Choose distinct primes $p_1, \ldots, p_k$ and let $A = p_1 \cdots p_k$. (Often $A = p_1 p_2$.) Then $\varphi(A) = (p_1 - 1) \cdots (p_k - 1)$. Choose any number e that is relatively prime to $\varphi(A)$. To determine exactly which numbers e are relatively prime requires us to factor $\varphi(A)$, but we don't need to do that. We just need to find some value of e, and we can do that by trial and error. Choose a value of e, apply Euclid's algorithm, and see if e is relatively prime to $\varphi(A)$. If so, fine. If not, choose a different value of e. Keep trying until we find one that works. We publicize e and A.

Suppose someone wants to send you a message. The sender breaks it down into blocks $b_1, \ldots, b_n$ with $0 \le b_i < A$ for each i. The sender enciphers each of the blocks by computing

$$c_i \equiv b_i^e \pmod{A}, \qquad 0 \le c_i < A,$$

and sends you the enciphered message $c_1, \ldots, c_n$.

You decode the message as follows. Since you know e and $\varphi(A)$, you can easily find positive integers s and t with

$$1 = se - t\varphi(A).$$

(Use Euclid's algorithm to find integers s_0 and t_0 with $1 = s_0 e - t_0 \varphi(A)$. Then s_0 and t_0 must be either both positive or both negative. If they are both positive, set $s = s_0$ and $t = t_0$. If they are both negative, note that also $1 = (s_0 + m\varphi(A))e - (t_0 + me)\varphi(A)$ and choose m large enough so that both $s_0 + m\varphi(A)$ and $t_0 + me$

are positive. Then set $s = s_0 + m\varphi(A)$ and $t = t_0 + me$.) Then in either case $se = t\varphi(A) + 1$ with s and t positive.

Now if $c \equiv b^e \pmod{A}$, then

$$c^s \equiv (b^e)^s \equiv b^{se} \equiv b^{t\varphi(A)+1} \equiv b \pmod{A}.$$

Thus you simply compute

$$b_i \equiv c_i^s \pmod{A}, \qquad 0 \le b_i < A,$$

and the decoded message is $b_1, \ldots, b_n$.

You can easily decipher the message because you know $\varphi(A)$, which you keep secret. The sender does not need to know $\varphi(A)$ to send the message. If this message is intercepted, the person who intercepts it needs to know $\varphi(A)$ in order to decipher it. In order to find $\varphi(A)$ this person needs to factor A. But, given the present state of mathematical knowledge and computer hardware, for A sufficiently large this is an intractable problem. (It is easy to write down an A large enough so that factoring A will take centuries.) But even for huge values of A, the basic computation you have to do is to find s and t, and you can do that quickly by using Euclid's algorithm.

Example 3.6.3. For purposes of illustration we will assume our messages consist only of letters in the English alphabet. We will represent these letters by $01 = A, \ldots, 26 = Z$. (Thus we will not worry about upper- versus lowercase letters, spaces between words, punctuation, etc.) Also, in this example we will break up our messages into 3-letter blocks.

Let $A = 379 \cdot 797 = 302063$, a product of two primes. Then $\varphi(A) = 378 \cdot 796 = 300888$. Choose $e = 13579$. Using Euclid's algorithm we compute that

$$300088 = 13579(22) + 2150$$
$$13579 = 2150(6) + 679$$
$$2150 = 679(3) + 113$$
$$679 = 113(6) + 1$$
$$113 = 1(113)$$

so 13579 and 300888 are relatively prime, and furthermore

$$1 = 679 + 113(-6)$$
$$= 679 + [2150 + 679(-3)](-6) = 2150(-6) + 679(19)$$
$$= 2150(-6) + [13579 + 2150(-6)](19) = 13579(19) + 2150(-120)$$
$$= 13579(19) + [30888 + 13579(-22)](-120) = 300888(-120)$$
$$\quad + 13579(2659).$$

Thus

$$1 = 13579(2659) - 300888(120),$$

so $s = 2659$.

Your secret agent wants to transmit to you the message "Jet fighter," encodes it as 100520 060907 082005 182727 (with the last block "padded" so that it has length 3), enciphers it by computing

$$100520^{13579} \equiv 186010 \ (\text{mod} \ 302063)$$

$$060907^{13579} \equiv 249959 \ (\text{mod} \ 302063)$$

$$082005^{13579} \equiv 186010 \ (\text{mod} \ 302063)$$

$$182727^{13579} \equiv 143782 \ (\text{mod} \ 302063)$$

and sends you 186010 249959 186010 143782.

You receive another message from your secret agent, which reads 022689 001784 246400. You decipher it by computing

$$022689^{2659} \equiv 192102 \ (\text{mod} \ 302063)$$

$$001784^{2659} \equiv 130118 \ (\text{mod} \ 302063)$$

$$246400^{2659} \equiv 091405 \ (\text{mod} \ 302063)$$

which you decode into "Submarine."   ◇

Remark 3.6.4. It may seem that evaluating x^a (mod A) is a very slow process for a large, but this can in fact be done very quickly. The key observation to make is that $x^2 = (x)^2$, $x^4 = (x^2)^2$, $x^8 = (x^4)^2$, etc. In other words, x^{2^n} can be evaluated by only n successive squarings. Then x^a can be evaluated quickly by expressing a as a sum of distinct powers of 2 and taking products. For example, calculation shows that $2659 = 2048 + 512 + 64 + 32 + 2 + 1$ (or, equivalently, that $2659 = 101001100011_2$), and then $x^{2569} \equiv x^{2048} x^{512} x^{64} x^{32} x^2 x$ (mod A). Observe also that the computation can be reduced (mod A) at every step in order to keep the numbers small.   ◇

Appendix A

Logical Equivalence of the Various Forms of Induction

In this appendix, we prove the logical equivalence of the various forms of induction.

Before doing so, we need to be clear about what it means for two methods of proof to be logically equivalent.

Suppose we have Method I, which requires us to have a set of Hypotheses that we call Hypotheses I, and gives us a conclusion that we call Conclusion I, and suppose similarly that we have Method II, with a set of hypotheses Hypotheses II and a conclusion Conclusion II.

Then our two methods are logically equivalent if we can use either one in place of the other. That is, they are equivalent if on the one hand, whenever we have Hypotheses II, we can use Method I to reach Conclusion II, and on the other hand, whenever we have Hypotheses I, we can use Method II to reach Conclusion I.

This statement goes in two directions. Let us break it up, as we will want to consider each direction separately. We shall say that Method A implies Method B if whenever we have Hypotheses B, we can use Method A to reach Conclusion B. Thus Methods I and II are equivalent if Method I implies Method II and also Method II implies Method I.

Theorem A.1.1. *The following are logically equivalent:*

(a) The principle of mathematical induction

(b) The principle of complete induction

(c) The well-ordering principle.

Proof. Since this theorem is a bit subtle, rather than prove it in a few big steps we shall prove it in more small steps. To this end we begin by showing that well-ordering is equivalent to two other proof methods. These do not have standard names, so we shall simply call them well-ordering(*) and well-ordering(**). (Well-ordering, well-ordering(*), and well-ordering(**) are all very similar, but of course not identical.)

We consider these one at a time.

Recall that the well-ordering principle is the axiom:

Axiom A.1.2. *Let T be the set of all positive integers. Then any nonempty subset S of T has a least element.*

We let well-ordering(*) be the axiom:

Axiom A.1.3. *Let T be the set of all positive integers. Then, for any positive integer n, any subset S of T that contains n has a least element.*

We observe that well-ordering and well-ordering(*) have different hypotheses but the same conclusion: The set S has a smallest element.

Let us begin by assuming that we have a set S and we are in a position to apply well-ordering(*). Then S contains some positive integer n. But that means that S is nonempty. Hence by well-ordering, S has a smallest element.

On the other hand, let us now assume that we have a set S and we are in a position to apply well-ordering. Then S is nonempty. But that simply means that S contains some positive integer n_1. But by well-ordering(*) any set that contains any positive integer n has a least element, so in particular we may choose $n = n_1$, and then conclude that S has a smallest element.

Thus we see that well-ordering and well-ordering(*) are equivalent.

We let well-ordering(**) be the axiom:

Axiom A.1.4. *Let T be the set of all positive integers. Then, for any positive integer n, any subset S of T that contains some integer less than or equal to n has a least element.*

We observe that well-ordering and well-ordering(**) have different hypotheses but the same conclusion: The set S has a smallest element.

Let us begin by assuming that we have a set S and we are in a position to apply well-ordering(**). Then S contains some positive integer less than or equal to n. But that means that S is nonempty. Hence by well-ordering, S has a smallest element.

On the other hand, let us now assume that we have a set S and we are in a position to apply well-ordering. Then S is nonempty. But that simply means that S contains some positive integer n_1. But by well-ordering(**) any set that contains a positive integer less than or equal to any positive integer n has a least element, so in particular we may choose $n = n_1$, and then conclude that S has a smallest element.

Thus we see that well-ordering and well-ordering(**) are equivalent.

Having introduced these two new proof methods, we now turn to considering complete induction and induction. Again, we consider these one at a time.

First we consider complete induction.

Claim: Complete induction implies well-ordering(*).

Let $P(n)$ be the statement that a subset S of T that contains the integer n has a smallest element. Then well-ordering(*) is just the claim that $P(n)$ is true for every positive integer n. We want to use complete induction to prove this.

The base case: Let $n = 1$. Then $P(1)$ is the statement that any subset of T that contains the number 1 has a smallest element. But since 1 is the smallest element of T, it is certainly the smallest element of S. In particular, S has a smallest element. Thus $P(1)$ is true.

The inductive step: Assume that $P(k)$ is true for $1 \leq k \leq n$ and consider $P(n+1)$. Let S be a subset of T that contains the integer $n+1$. There are two cases:

(a) S does not contain any integer k with $1 \leq k \leq n$. Then $n+1$ is certainly the smallest element of S. In particular, S has a smallest element.

(b) S contains some integer k with $1 \leq k \leq n$. By the inductive hypothesis, $P(k)$ is true, so S has a smallest element.

Thus in either case $P(n+1)$ is true. Hence, by complete induction, $P(n)$ is true for every positive integer n, i.e., well-ordering(*) is true.

Claim: Well-ordering implies complete induction.

Let $P(n)$ be a proposition about the positive integer n. We assume the hypotheses of complete induction hold, and we want to use well-ordering to derive the conclusion of complete induction, which is that $P(n)$ is true for every n.

We prove this by contradiction. Suppose that it is not the case that $P(n)$ is true for every n. Let S be the set of positive integers n for which $P(n)$ is false. Then S is nonempty.

By well-ordering, S has a smallest element. Call it n_1. What can n_1 be?

(a) We cannot have $n_1 = 1$, as by the hypotheses of complete induction, $P(1)$ is true.

(b) Thus $n_1 > 1$. Write $n_1 = n+1$. We cannot have that $P(k)$ is true for all $1 \leq k \leq n$, as by the hypotheses of complete induction, that would imply that $P(n+1)$ would be true.

Thus it must be the case that $P(k)$ is false for some k with $1 \leq k \leq n$. But that means k is in S, which is impossible, as $k < n_1$ and n_1 is the smallest element of S.

Thus at this point we have that complete induction implies well-ordering(*), which is equivalent to well-ordering, and that well-ordering implies complete induction, and we conclude that complete induction and well-ordering are equivalent.

Now we consider induction.

Claim: Induction implies well-ordering(**).
Let $P(n)$ be the statement that a subset S of T that contains some integer less than or equal to the integer n has a smallest element. Then well-ordering(**) is just the claim that $P(n)$ is true for every positive integer n. We want to use induction to prove this.

The base case: Let $n = 1$. Then $P(1)$ is the statement that any subset of T that contains some integer less than or equal to 1 has a smallest element. But the only positive integer less than or equal to 1 is 1 itself. Thus $P(1)$ is just the statement that any subset of T that contains the number 1 has a smallest element. But since 1 is the smallest element of T, it is certainly the smallest element of S. In particular, S has a smallest element. Thus $P(1)$ is true.

The inductive step: Assume that $P(k)$ is true for $1 \le k \le n$ and consider $P(n+1)$. Let S be a subset of T that contains some integer less than or equal to $n+1$. There are two cases:

(a) S does not contain any positive integer k less than $n+1$. Then $n+1$ is certainly the smallest element of S. In particular, S has a smallest element.

(b) S contains some positive integer k less than $n+1$. Then k is less than or equal to n. By the inductive hypothesis, $P(n)$ is true, so S has a smallest element.

Thus in either case $P(n+1)$ is true. Hence, by induction, $P(n)$ is true for every positive integer n, i.e., well-ordering(**) is true.

Claim: Well-ordering implies induction.
Let $P(n)$ be a proposition about the positive integer n. We assume the hypotheses of complete induction hold, and we want to use well-ordering to derive the conclusion of complete induction, which is that $P(n)$ is true for every n.

We prove this by contradiction. Suppose that it is not the case that $P(n)$ is true for every n. Let S be the set of positive integers n for which $P(n)$ is false. Then S is nonempty.

By well-ordering, S has a smallest element. Call it n_1. What can n_1 be?

(a) We cannot have $n_1 = 1$, as by the hypotheses of complete induction, $P(1)$ is true.

(b) Thus $n_1 > 1$. Write $n_1 = n+1$. We cannot have that $P(n)$ is true, as by the hypotheses of complete induction, that would imply that $P(n+1)$ would be true.

Thus it must be the case that $P(n)$ is false. But that means n is in S, which is impossible, as $n < n_1$ and n_1 is the smallest element of S.

Thus at this point we have that induction implies well-ordering(**), which is equivalent to well-ordering, and that well-ordering implies induction, and we conclude that induction and well-ordering are equivalent.

Finally, since complete induction is equivalent to well-ordering, and induction is equivalent to well-ordering, we conclude that complete induction and induction are equivalent to each other. □

(This proof evidently contains a certain amount of repetition. But that is deliberate. Our intention was to produce the simplest proof, not the shortest one.)

Index